Current Topics in Microbiology 250 and Immunology

Springer
Berlin
Heidelberg
New York
Barcelona
Hong Kong
London
Milan
Paris
Singapore
Tokyo

Clostridium difficile

Edited by K. Aktories and T.D. Wilkins

With 20 Figures and 11 Tables

 Springer

Professor Dr.Dr. KLAUS AKTORIES
Albert-Ludwigs-Universität Freiburg
Institut für Pharmakologie und Toxikologie
Hermann-Herder-Str. 5
79104 Freiburg i.Br.
Germany
E-mail: aktories@uni-freiburg.de

Dr. TRACY D. WILKINS
Director of Biotechnology
Virginia Tech
Fralin Biotechnology Center
West Campus Drive, MC 0346
Blacksburg, VA 24061
USA
E-mail: tracyw@vt.edu

Cover Illustration: Biopsy specimen from a patient with pseudo-membranous colitis induced by Clostridium difficile. *Focal superficial necrosis of the colonic mucosa with a typical mushroom-like desposition of fibrin with a few granulocytes. [Courtesy of Prof. Dr. E.-H. Schaefer, Pathologisches Institut (Ludwig-Aschoff-Haus) der Universität Freiburg, Germany]*

ISSN 0070-217X
ISBN 3-540-67291-5 Springer-Verlag Berlin Heidelberg New York

Springer-Verlag is a company in the BertelsmannSpringer publishing group
© Springer-Verlag Berlin Heidelberg 2000
Library of Congress Catalog Card Number 15-12910
Printed in Germany

Cover Design: *design & production GmbH*, Heidelberg
Typesetting: Scientific Publishing Services (P) Ltd, Madras
Production Editor: Angélique Gcouta
Printed on acid-free paper   SPIN: 10664872   27/3020GC 5 4 3 2 1 0

Preface

Since *Clostridium difficile* was recognized as a major pathogen for a wide range of enteric diseases including antibiotic-associated diarrhea and pseudomembranous colitis, it has become a major focus of intensive research. Scientific contributions on *Clostridium difficile* come from remarkably diverse fields including clinical and epidemiological research, microbiology, molecular biology, pharmacology, and cell biology. These concerted efforts have resulted in significant progress in the understanding of the role and action of *Clostridium difficile* as a pathogen. Progress in the molecular biology and toxinology of *Clostridium difficile* has been especially fruitful. Characterization of the mode of action of *Clostridium difficile* toxins A and B, which are the crucial virulence factors of the pathogens, has opened up a scientific treasure trove. The toxins have become the prototypes of a new family of clostridial cytotoxins that modify eukaryotic targets by glucosylation. The specific targets for the toxins are Rho GTPases, which regulate the actin cytoskeleton and act as molecular switches to control a large array of signaling processes. The unique activity and specificity of these toxins make them useful pharmacological and cell biological tools for studying signaling pathways in which Rho GTPases are involved.

This volume gives a review and update of recent developments and progress in this exciting field of research. We hope that the various contributions from different scientific fields will inspire further studies on *Clostridium difficile* and clostridial toxins.

KLAUS AKTORIES and TRACY D. WILKINS

List of Contents

List of Contributors

(Their addresses can be found at the beginning of their respective chapters.)

AKTORIES, K.   55

BOQUET, P.   97

BORRIELLO, S.P.   1

BRAZIER, J.S.   1

CHAVES-OLARTE, E.   85

FARRELL, R.J.   109

GERDING, D.N.   127

HOFMANN, F.   55

JUST, I.   55, 97

LAMONT, J.T.   109

MONCRIEF, J.S.   35

THELESTAM, M.   85

WILKINS, T.D.   35

Microbiology, Epidemiology and Diagnosis of *Clostridium difficile* Infection

J.S. Brazier[1] and S.P. Borriello[2]

1 Introduction

Clostridium difficile made its first appearance in the literature when HALL and O'TOOLE (1935) described *Bacillus difficilis* as part of the bacterial flora of the

---

[1] Anaerobe Reference Unit, Public Health Laboratory, University Hospital of Wales, Heath Park, Cardiff CF4 4XW, UK
[2] Central Public Health Laboratory, 61 Colindale Avenue, London NW9 5HT, UK

meconium and faeces of infants. Although they postulated that toxins from certain strains, when liberated in the infant gut, might play a role in conditions such as the formation of occult blood and febrile convulsions of the newborn, it was not until 1969 that the first real clue to the pathogenic potential of this organism to mammals in the absence of competing colonic microbiota was provided. In experiments on germ-free rats, it was noted that mono-contamination with *C. difficile* often led to development of transient diarrhoea, which occasionally caused death (HAMMARSTROM et al. 1969). The significance of this finding remained unappreciated until the mid 1970s.

As the organism was an obligately anaerobic Gram-positive spore-bearing bacillus, it was subsequently classified as belonging to the genus *Clostridium*, and for the next few decades only made fleeting appearances in the literature. McBEE (1960) isolated *C. difficile* from the gut of a Weddell seal, and SMITH and KING (1962), looking specifically for reports of *C. difficile* in human infections, noted eight incidents of extra-intestinal infection in which they concluded it was not playing a pathogenic role. The second report of an animal origin was that of STEVENSON (1966) who found *C. difficile* in the gut contents of a desert locust.

1.1 *C. difficile* as a Pathogen to Man

Three independent studies, concluded in 1974, provided the platform from which *C. difficile* was shown to be an important cause of disease in man, ironically, mainly at the opposite end of our life span than was envisaged by Hall and O'Toole. In the United States, GREEN (1974) described a cytotoxin that was present in the stools of guinea pigs which developed gut disease after receiving penicillin. TEDESCO et al. (1974), also in the United States, found a significant association between patients receiving clindamycin and the development of pseudomembranous colitis (PMC). At this stage neither knew the aetiology of their observations. Meanwhile in Leeds, England, HAFIZ (1974) studying under Professor C.L. Oakley, completed his PhD thesis on *C. difficile* and its toxicity. He was unaware, however, that the organism he was studying was responsible for the toxic effects noted by workers in the United States. These independent publications were a catalyst to studies by BARTLETT et al. (1977) who described a clindamycin-induced colitis in hamsters. They isolated an unidentified *Clostridium* species from the hamsters, which was eventually confirmed as *C. difficile* and was implicated as the cause. LARSON et al. (1977) demonstrated that a cytotoxin could also be detected in the stools of five out of six patients with histologically proven PMC. There soon followed several studies which provided confirmation of the causal association of *C. difficile* with gut infection in man (BARTLETT et al. 1978; GEORGE et al. 1978).

1.2 Microbiology

C. difficile is an anaerobic spore-bearing bacillus, 3–5µm in length, that stains predominantly Gram-positive, although older colonies may exhibit marked Gram stain variability. Sporulation is most noticeable on agar cultures that have reached a stationary or decline phase after >72h incubation. Of note is the fact that *C. difficile* does not sporulate on most formulations of *C. difficile* selective media. Colonial morphology can be quite variable; typically at 24–48h on blood-based media, colonies are 3–5mm in diameter with an irregular, lobate or rhizoidal edge, grey, opaque and non-haemolytic, although some strains have a greenish appearance due to an alpha-type haemolysis on blood agar. After 48–72h incubation, colonies may develop a light grey or whitish centre; a factor associated with sporulation. This centre is sometimes raised with an inner concentric margin giving almost a "fried-egg" appearance. Smooth forms may also occur, especially on selective agars. Dwarf colonies are common and represent colonies formed from inoculated spores that have subsequently germinated (J.S. Brazier, personal observation).

1.3 Ecology of *C. difficile*

C. difficile is normally a harmless environmental organism. As is the case with some other bacterial species, it is man's intervention which has facilitated the conditions whereby it may cause significant human morbidity and mortality. It is the compromise of resistance to colonisation afforded by the normal gut flora, usually by antimicrobial agents, that facilitates infection by *C. difficile*. As ingestion is required, a source of the organism must exist and the environment is one possibility. The first study to screen specifically for *C. difficile* in the inanimate environment was that of HAFIZ (1974) who isolated the organism from soil and sand in Pakistan, from river bank mud, from hay and from the dung of a camel, a donkey, a cow and a horse. Others have also found the organism in soil (BLAWAT and CHYLINSKI 1958), but some only found it in sites related geographically or temporally to human sewage or patients with active *C. difficile* infection, while others failed completely to isolate it from soil samples (KIM et al. 1981; RILEY 1994). This suggests either irregular environmental distribution, or more likely, that the results were affected by variable sampling methodology. To date, the largest study of the environment as a source of *C. difficile* was that of AL SAIF and BRAZIER (1996) who examined 2580 environmental samples in the Cardiff area of South Wales, United Kingdom. Their survey included samples of river and sea water, swimming-pool water, raw vegetables, soil, farm and pet animal faeces, in addition to general surfaces in homes, hospitals and veterinary clinics which were examined using a methodology designed to maximise spore recovery. Their results showed an overall positivity rate of 7.1% with largely expected results for soil (21%), pets (7%) and hospitals (20%). The most significant findings were that 87.5% of river waters and 46.7% of lake waters were positive for *C. difficile*, as well as 50% of swimming pools, 2.4% of raw vegetables and 2.2% of surfaces in private homes. Here, too, methodology may have been a key factor, as a protocol of 5 days incubation using a medium designed to enhance spore germination was employed which often resulted in isolation of *C. difficile* after 3–5 days incubation from cultures which were negative at 48h. Such environmental data may explain how the population in the community could be exposed to *C. difficile* and henceforth carry the organism into hospitals.

1.4 *C. difficile* Carriage in Animals

There is no direct evidence that *C. difficile* infection in man has zoonotic origins despite the fact that numerous animal reservoirs have been recognised and a great variety of wild, farmed and domesticated species may carry the organism (BORRI-ELLO et al. 1983; RILEY et al. 1991; LEVETT 1986). Carriage in animals in the human food chain was investigated by AL SAIF and BRAZIER (1996) who sampled 524 faecal or gut contents from cattle, sheep, pigs, fish and poultry, with isolation rates ranging from 0 to 1.6%, concluding that neither meat or fish was an important reservoir. Disease due to *C. difficile* infection in animals has been recognised ini-tially in hamsters treated with clindamycin (BARTLETT et al. 1977). Horses, too, can suffer *C. difficile* infection as shown in a study where nine of 10 hospitalised horses developed diarrhoea after becoming colonised with toxigenic strains within a 2-day period (MADWELL et al. 1995). Enteritis and enterotoxaemia due to *C. difficile* in captive ostriches (FRAZIER et al. 1993) has also been reported. Chronic diarrhoea in dogs can apparently be due to *C. difficile* according to BERRY and LEVETT (1986). However, STRUBLE et al. (1994) could find no direct association with toxin positive isolates and dogs with diarrhoea in a veterinary teaching hospital, while BORRIELLO et al. (1983) found no association between carriage and diarrhoea in household pets. *C. difficile* infection in mammals usually requires the same disturbance of gut flora by antibiotics as in man.

1.5 *C. difficile* Carriage in Humans

Asymptomatic human intestinal carriage of *C. difficile* has been studied in various population groups. The highest reported healthy adult carriage rates come from Japan where NAKAMURA et al. (1981) found a 15% carriage rate. In contrast, LARSON et al. (1978) in England found no healthy carriers in a small study of 41 samples from 11 volunteers. A Swedish study (ARONSSON et al. 1985) of 594 vol-unteers found a carriage rate of 1.9% and PHILLIPS and ROGERS (1981) reported 2% isolation rate in 100 random stool samples that were negative for other pathogens. However, many factors may influence the results of such studies; for instance, the size of the study, the ethnic origin of the study group, and the fact that subjects should not have taken antibiotics for at least a month before sampling may have been overlooked.

There is more concordance in the literature on the high rates of carriage in the faeces of healthy neonates and infants, as first shown in the seminal study of HALL and O'TOOLE (1935). A Swedish study (HOLST et al. 1981) found *C. difficile* in 60% of 218 healthy neonates, peaking at a rate of 64% carriage in the age group 1–8 months. LARSON et al. (1982) demonstrated that acquisition of the organism varied within the same institution. In their study, the same cultural methods were used on babies from three neonatal wards and carriage rates varied from 2% to 52%. DONTA and MYERS (1982) reported that 10.5% of 105 healthy newborns surveyed were culture positive and that 55% (28/51) of babies on a neonatal intensive care unit had toxin positive stools. BOLTON et al. (1984) sampled 66 babies in the first week of life, born in a single maternity unit, finding *C. difficile* in 31 (47%). EL MOHANDES et al. (1993) investigated 50 pre-term babies housed in an intensive care facility over the first 34 days of life sampling at weekly intervals. They found a 15% isolation rate during week 1, rising to 33% at week 2, and this level was maintained for the remaining period of study. Isolates were tested for cytotoxin (toxin B) production, and toxigenic strains accounted for 71–100% of isolates over the sampling period. Why infants remain unaffected by the toxins is a topic of debate. It is possible that they lack the receptors for either toxin, or that the receptors are masked. However, there is as yet no experimental proof of this.

Studies by BOLTON et al. (1984) and SHERERETZ and SARUBBI (1982) suggested nosocomial acquisition was the most probable route by which babies became colonised as opposed to vertical transmission suggested by others. Credence to this theory was provided by LARSON et al. (1982) and MALAMOU-LADAS et al. (1983), the latter of which isolated *C. difficile* from 3.2% of inanimate objects sampled in a neonatal unit. Conversely, vaginal delivery and breast-feeding were associated with cytotoxin positive stool assay in one study (DONTA and MYERS 1982). However, these findings have not been supported by others (EL MOHANDES et al. 1993); the suggested reason being the failure to use enrichment methods. Genito-urinary tract carriage of *C. difficile* is itself a contentious issue. HAFIZ et al. (1975) reported a 71.7% isolation rate from the vaginas of females attending a genito-urinary (GU) clinic, with a much lower carriage rate of 18.1% in women attending a family planning clinic. In male patients attending a GU clinic, they reported a urethral carriage rate of 100% in 42 patients suffering from non-specific urethritis. A control group of 50 men attending a urology clinic had a zero carriage rate and the authors therefore concluded that *C. difficile* was pathogenic in this situation. These findings have not been substantiated by others, however, and are generally regarded with scepticism.

Despite all this information, it is not known if gut carriage of *C. difficile* is a permanent or temporary state. A longitudinal study using molecular methods such as polymerase chain reaction (PCR) to detect low numbers is required to elucidate if carriage of *C. difficile* in the healthy gut of adults is primarily transient with any isolates merely "en passant" at the time of sampling.

1.6 *C. difficile* Infection in the Community

As *C. difficile* infection is primarily a nosocomial problem, relatively little attention has been paid to the degree of *C. difficile* diarrhoea in the community. Much of the published work in this field comes from Western Australia (RILEY et al. 1991,

1995). In the first of these studies, 288 stool samples from general practice clinics were examined for cytotoxin and presence of the organism; 5.5% of stools were positive for either, making it more common than *Campylobacter* or *Salmonella* (RILEY et al. 1991). The second study (RILEY et al. 1995) investigated 580 faecal samples yielding 75 positives from 61 patients (10.7%) which made it the second most common pathogen after *Campylobacter*. In the United States, HIRSCHORN et al. (1994) investigated the level of *C. difficile* infection in the community by a retrospective cohort study of members of a health maintenance organization. Their 51 cases corresponded to an incidence level of 7.7 cases per 100,000 person years. 65% of cases had occurred within 42 days of receiving antibiotics and there was a significant association with increased age ($p < 0.001$). Certain antibiotics were more associated with symptoms in age-adjusted antibiotic-specific attack rates. Symptoms were at least tenfold higher in patients receiving nitrofurantoin, cefuroxime, cephalexin with dicloxacillin, ampicillin/clavulanate with cefaclor, and ampicillin/clavulanate with cefuroxime than in those receiving ampicillin or amoxycillin alone. A large-scale community-based study in England (in press), recently investigated the level and causes of diarrhoeal disease in patients in the community attending their general practitioners. *C. difficile* cytotoxin was found in 1.7% of diarrhoeal cases presented to general practitioners and there was a significant correlation with prior exposure to antibiotics. This evidence suggests that *C. difficile* may cause diarrhoeal disease in non-hospitalised patients and further study is warranted in this area.

1.7 Extra-Intestinal Infections with *C. difficile*

Although the gastro-intestinal tract of man is the main focus of infection, *C. difficile* may occur occasionally in clinical material other than faeces. Seventeen cases were reviewed by FELDMAN et al. (1995) and BYL et al. (1996) similarly described three cases in which severe underlying conditions, such as malignancy, were believed to have predisposed two of these cases to *C. difficile* bacteraemia and an abscess in the pouch of Douglas. Reactive arthritis associated with *C. difficile* in a patient with Reiter's syndrome has also been reported (McCLUSKEY et al. 1982). Recently, one of the authors (JSB) has investigated two cases of sub-phrenic abscesses from which pure cultures of *C. difficile* (one toxigenic, the other non-toxigenic) were isolated, and also a case of *C. difficile* bacteraemia.

1.8 Risk Factors and Diagnostic Indicators of *C. difficile* Infection

Four factors are believed to dictate the risk of *C. difficile* infection in man. These are: distortion of stable colonic flora by antibiotics, exposure to and acquisition of a toxigenic strain, colonisation and toxin production, and age-related susceptibility. Previously, GERDING and BRAZIER (1993) had reviewed the criteria for optimal diagnosis of infections due to *C. difficile*. Their key elements were: the patient's clinical history and symptoms, endoscopic evidence of a colonic pseudomembrane, laboratory evidence of *C. difficile* in the stool, and laboratory evidence for toxin(s) in the stool. These key elements in relation to both clinical and laboratory diagnosis of *C. difficile* infection are discussed below.

There are a number of risk- and host-related factors which should alert both the clinician and the laboratory to the likelihood of *C. difficile* infection in a given hospitalised patient, with old age as a major risk factor. *C. difficile* infection is generally more of a problem in patients over 65 years of age (McFARLAND et al. 1990). In 1998, the Communicable Disease Surveillance Centre (CDSC) of the Public Health Laboratory Service (PHLS) recorded that stools from patients in this age group accounted for approximately 97% of all positive laboratory reports in England and Wales (ANONYMOUS 1998). A history of antimicrobial therapy or prophylaxis in a patient with diarrhoea is strongly indicative of antibiotic-associated diarrhoea with *C. difficile* as a possible cause. However, antibiotics alone should not be a pre-requisite for laboratory investigation, since other medications such as cytotoxic drugs, antacids, stool softeners and laxatives may also trigger *C. difficile* infection. Certain procedures such as naso-gastric intubation, enemas and other intensive care procedures may also predispose to *C. difficile* infection (SPENCER 1998). While some antimicrobial agents are more often associated with *C. difficile* infection (e.g. oral broad spectrum β-lactams and third-generation cephalosporins), investigation for *C. difficile* infection should not be ruled out on the basis of the patient having received antimicrobials with a low association with *C. difficile* infection (e.g. oral aminoglycosides and some of the quinolines). KATZ et al. (1997) reported that in patients with a combination of antibiotic therapy, those with significant diarrhoea and abdominal pain were the group most likely to have a positive cytotoxin assay.

Diarrhoea is the most common clinical symptom, but this can be difficult to define. A general definition is more than three watery, loose, or unformed stools (defined as those which take the shape of the container into which they have been voided) per day for at least 2 days. It is important also, however, to consider the possibility of a non-infectious cause of diarrhoea before considering an infectious cause such as *C. difficile*. Ward nurses are often the first to notice a change in a patient's bowel motions, and there is much anecdotal evidence from experienced nurses suggesting that a distinctive odour is associated with the stool of a patient with *C. difficile* infection. An experienced nurse may therefore play a vital role in early diagnosis, and this in turn may reduce the likelihood of cross-infection in the ward. Patients with diarrhoea caused by *C. difficile* may suffer excess protein loss whereby occult protein is lost into the faeces. In severe cases this may contribute to the malnutrition of an already debilitated and often elderly patient. This protein loss results in hypoalbuminaemia, reduced serum levels of cholesterol and transferrin and elevated levels of faecal α_1-antitrypsin. These parameters are within normal limits in patients who are asymptomatic carriers of *C. difficile*, therefore faecal α_1-antitrypsin could be useful for diagnosis (DANSINGER et al. 1996).

Diarrhoea may be absent due to an ileus, and severely ill patients may present with hyper-pyrexia, moderate to severe abdominal pain, bowel perforation, toxic megacolon or PMC. Indeed, some diagnoses of *C. difficile* infection have only been made histologically following post-mortem examination. These patients frequently require endoscopic examination. Endoscopic diagnosis may be particularly helpful for patients with particularly severe abdominal pain when exploratory surgery is being considered. If classical pseudomembranes are seen by endoscopy, surgery may be avoided; a negative finding, however, does not rule out *C. difficile* infection, since this procedure has a limited sensitivity, especially if the shorter rigid proctoscope is used rather than the longer flexible fibre optic colonoscope (GERDING and BRAZIER 1993).

Once the clinician suspects antibiotic-associated diarrhoea (AAD) or more serious *C. difficile* infection, laboratory confirmation entails submission of a faecal specimen to the hospital microbiology laboratory.

2 Laboratory Diagnosis of *C. difficile* Infection

Optimal laboratory investigations are performed on a freshly taken faecal specimen which should be submitted immediately to the laboratory. Storage at ambient temperatures for prolonged periods leads to possible denaturation of faecal toxin. BOWMAN and RILEY (1986) reported a 100-fold decrease in cytotoxin titre of specimens stored at 22°C, and BRAZIER (1993) reported complete inactivation of cytotoxin in approximately 20% of stool samples sent through the mail. BORRIELLO et al. (1992) stored a toxin positive stool at 4°C and tested it for the presence of toxin A and cytotoxin periodically. After 44 days both were still detected, although after 52 days the immunoassay for toxin A became negative. Therefore, whenever it is not possible to examine freshly voided stools, specimens should be kept at 4°C or frozen at −20°C. Culture of *C. difficile* from stools should be largely unaffected by ambient storage since the organism sporulates readily; there is, however, a paucity of information on this subject in the literature. If there is any delay in processing faecal specimens it is advisable to store specimens at lower temperatures since *C. difficile* will survive in faeces for many months at 4°C and for over a decade frozen at −70°C (J.S. Brazier, personal observation).

The Department of Health/Public Health Laboratory Service report (1994) stated that formed stools should not be examined for *C. difficile* since free toxin is not usually found in solid stools, and also because toxigenic and non-toxigenic strains may be carried asymptomatically.

2.1 Laboratory and Bedside Criteria for Laboratory Investigation of *C. difficile*

There are widely varying criteria by which a microbiology laboratory may, or may not, choose to examine a stool specimen for *C. difficile*. A questionnaire on selection criteria for *C. difficile* examination was conducted by Wilcox and Smyth (1998). Of 104 respondents, only 47% tested for *C. difficile* if specifically requested, 19% if antibiotic use was stated, 15% tested all diarrhoeal specimens and 14% selected in-patients only. Bowman and Riley (1988) maintain that infectious diarrhoea in a hospitalised patient is more likely to be due to *C. difficile* than any other enteric pathogen, and therefore include investigations for it as routine. As costs are an important factor, however, some laboratories may look to save money by being very selective in this area. Clinicians too may also come under pressure to restrict their requests for laboratory tests. Manabe et al. (1995) studied 268 consecutive patients who were investigated for *C. difficile* infection, in an attempt to establish the optimum number of stool samples needed to reasonably exclude the diagnosis of *C. difficile* infection. They reported a 97% negative predictive value for the first stool examination, and suggested that clinicians should take note of readily available clinical signs and the results of biochemistry investigations (as mentioned previously) to decide when it is most appropriate to order specific tests for *C. difficile*. Renshaw et al. (1996) looked specifically at the value of repeated requests for stool cytotoxin and concluded that little useful information was gained ordering repeat assays within a 7-day period of the initial request. Although this analysis was not restricted to requests on diarrhoeal stools, their survey found repeat requests accounted for 36% of the total assays performed. Repeat testing during and after treatment is of no clinical value unless relapse of symptoms occurs. Stools from community and general practice patients are often not tested for *C. difficile* by laboratories despite evidence that infection may occur in non-hospitalised patients (Riley et al. 1991).

3 Detection of *C. difficile* or its Metabolic Products

3.1 Microscopy

Microscopical examination of the stool specimen for the presence of leukocytes and erythrocytes has been proposed as an indicator of *C. difficile* infection (Bowman and Riley 1988). The value of Gram staining of faeces for the presence of polymorphonuclear leukocytes and clostridia was examined by Shanholtzer et al. (1983) but this procedure is very non-specific and has little to recommend it. Fluorescence microscopy on faecal smears using labelled antibodies to *C. difficile* cell wall antigens (Wilson et al. 1982) gave 93% correlation with cytotoxin and culture results, but 62% of normal healthy adult stool samples also gave a positive reaction by this method. An explanation could be the shared surface antigens of *C. difficile* with normal gut flora such as *C. bifermentans/sordellii*, which had been reported previously (Poxton and Byrne 1981). Neither microscopical approach has proved sufficiently sensitive or specific enough for use as the sole diagnostic test for *C. difficile* infection.

3.2 Faecal Volatile Fatty Acid Analysis

Metabolites in faeces indicative of *C. difficile* such as isovaleric acid, isocaproic acid, or *p*-cresol as detected by gas–liquid chromatography (GLC) has been evaluated by several workers. POTVLIEGE et al. (1981) reported 61% correlation between GLC results and cytotoxin plus culture. PEPERSACK et al. (1983) came to a similar conclusion, and suggested that GLC was an excellent screening method for excluding *C. difficile* infection. LEVETT (1984), however, reported that only 41% of culture positive stools contained detectable isocaproic acid or *p*-cresol, but that 85% contained isovaleric acid. Sadly this increased sensitivity was accompanied by a loss of specificity since many other gut anaerobes produce this metabolite. Frequency-pulsed electron capture GLC was applied to analysis of stools from a range of diarrhoeal syndromes (BROOKS et al. 1984). Isocaproic acid was found to be a good marker for the presence of *C. difficile*, being present in all cases that were culture positive and absent in all culture negative specimens; *p*-cresol was not specific for *C. difficile*. The complexity of this method has ruled against its adoption in routine laboratories.

3.3 Cultural Methods

Diagnosis of bacterial diarrhoeal diseases has traditionally centered upon cultural techniques to isolate the implicated pathogen. Compared to other bacterial gut pathogens, *C. difficile* is probably unique in that the main virulence factor (toxin production) of this bacterium was demonstrated before the pathogen producing the toxins was identified and optimal cultural procedures established.

HAFIZ and OAKLEY (1976) were the first to attempt selective isolation of *C. difficile*. This preceded knowledge of its involvement in gut disease. Their selective approach was based on the finding of others that *C. difficile* could deaminate the amino acid phenylalanine to form *p*-hydroxyphenylacetic acid and then decarboxylate this to form *p*-cresol (ELSDEN et al. 1976). They reported that incorporation of 0.2% *p*-cresol in a broth medium enabled isolation of *C. difficile* in pure culture from infant faeces, horse, camel and donkey dung, sand and mud. This method has not been generally adopted, however. Selective agars based on neomycin with sodium azide and kanamycin or clindamycin were tried subsequently with some success, but the first significant advance came with the development of cycloserine-cefoxitin fructose agar (CCFA) by GEORGE et al. (1979). The selective agents were used at concentrations of 500mg/l and 16mg/l respectively, and the medium also incorporated added carbohydrate in the form of fructose and neutral red as a pH indicator in an egg-yolk agar base. This effective formulation was used for cultural studies on *C. difficile* in many centres, although several modifications in both agar and broth formulae were published subsequently. WILLEY and BARTLETT (1979) dispensed with the fructose and egg-yolk base and reduced the concentrations of the selective agents to 250mg/l and 10mg/l respectively. This approach was later verified by LEVETT (1985) who reported increased isolation with half of the original strength of agents described by GEORGE et al. (1979). PHILLIPS and ROGERS (1981) incorporated an intermediate metabolite of *p*-cresol mentioned by Hafiz (*p*-hydroxyphenylacetic acid) as an aid to presumptive identification because of its distinctive odour and the possibility of demonstration of this metabolite by GLC. CARROLL et al. (1983) added gentamicin as a third selective agent in a broth formulation reporting an improved positivity rate of 13% isolation of *C. difficile* from 267 diarrhoeal stools examined by this method. Other workers chose to abandon the cefoxitin half of the selective agents, and substituted mannitol for fructose. This medium was called cefoxitin mannitol agar (CMA, or CMBA if supplemented with blood), and was compared to a modified CCFA by BARTLEY and DOWELL (1991) who tested 105 culture positive stools on each medium. They reported a 76% recovery rate with CMBA, 65% with CMB and only 36% or 42% recovery with CCFA depending on the commercial supplier of the medium. Alternative selective agents have also been tried. ASPINALL and HUTCHINSON (1992) used a combination of moxalactam (32mg/l) and norfloxacin (12mg/l) and compared it to CCFA on 832 faecal samples from hospitalised patients with diarrhoea. They reported a 20% higher yield and a 30% reduction in contaminating commensals with the newer medium, but this has not been substantiated by others.

Various groups have reported on the efficacy of alcohol-shock treatment of stool specimens to select for *C. difficile*. BORRIELLO and HONOUR (1981) mixed equal volumes of stool and ethanol and subcultured them on non-selective brain heart infusion agar after leaving to stand for 1h. Compared to direct inoculation on CCFA agar, this method was marginally less effective; however, BARTLEY and DOWELL (1991) found alcohol shock to be superior to direct culture on a range of selective agars. It is the author's (JSB) experience that the selection of spores by alcohol shock greatly diminishes competing flora which enhances both the isolation and easier recognition of *C. difficile*. The addition of bile salts such as sodium cholate or taurocholate to a medium is believed to enhance the recovery of spores by inducing the germination of spores from environmental samples or faeces after alcohol shock. BUGGY et al. (1983) reported increased recovery from hospital environmental sites using media supplemented with sodium taurocholate compared to CCFA. It is stressed that pure grade taurocholate must be used, however, since some unrefined desoxycholate salts may inhibit cell multiplication. The sodium salt of cholic acid is just as effective as pure taurocholate in stimulating spore germination and is one-tenth of the price. A medium incorporating ingredients for optimum isolation and recognition of *C. difficile* has been formulated and is commercially available (BRAZIER 1993). An important point regarding the viability of *C. difficile* colonies is that the organism does not readily sporulate on agars containing the selective agents. Colonies will therefore become non-viable if plates are left in air for prolonged periods. On non-selective agars, colonies will usually sporulate heavily after 72h incubation and hence survive prolonged exposure to air.

It is generally accepted that in cases of symptomatic disease, enrichment culture is unnecessary. However, for some environmental investigations, or surveys of asymptomatic carriage rates, selective enrichment may be useful and this method has been tried by various workers. A broth containing the original concentrations of cycloserine and cefoxitin was used by BUCHANAN (1984). He reported superior isolation rates from 401 stool samples compared to CCFA ($p < 0.001$) and implied that this method would be a valuable tool in epidemiological surveys. CLABOTS et al. (1991), however, found contact plates of CCFA to be superior to a broth formulation of the same ingredients in the recovery of *C. difficile* from environmental surfaces in a hospital ward.

3.4 *C. difficile* in Specimens Other than Faeces

Hafiz et al.'s report of *C. difficile* in the urethra of males with non-specific urethritis, and in the vaginas of 72% of women attending a genito-urinary clinic remains controversial and unsubstantiated (1975). Concerns about these uro-genital isolates was reported by BARTLETT (1988) which resulted in cultures being exchanged which he confirmed as *C. difficile*. Despite this alleged verification, the role and presence of *C. difficile* in genital specimens remains uncertain and it is not common practice to investigate specifically for it in material from these sites. *C. difficile* may occur in a variety of clinical sites such as wound infections (SMITH and KING 1962) but is usually part of a mixed flora and of doubtful significance. It is not necessary to select specifically for *C. difficile* in such specimens as it will grow on routine anaerobic selective agars.

3.5 Confirmation of *C. difficile*

Colonies of putative *C. difficile* can be recognised by a few simple tests and it is not cost-effective or necessary to perform a large range of phenotypic tests for confirmation on a routine basis. As described above, colonial morphology can be quite variable depending on the media constituents and length of incubation, and hence to the inexperienced worker is not a good marker. However, the odour associated with *C. difficile* colonies is very distinctive and has been likened by many to that of elephant or horse manure. Once this odour has been recognised it becomes a quick and easy aid to identification. Another characteristic is the ability of colonies to fluoresce a yellow-green, or chartreuse colour under long wave (365nm) ultra-violet illumination. This property is not unique to *C. difficile*, however, and as colonies age and sporulate they tend to become white and the fluorescence fades. It is also important to note that colonial fluorescence cannot be screened for on original formulations of CCFA as the neutral red pH indicator, auto-fluoresces (KAUFFMAN and WEAVER 1960). Despite this, when fluorescence is coupled to 48h growth on a blood-based primary isolation selective medium it is a very useful and recommended approach for detection of *C. difficile*. Another quick confirmatory test is the somatic antigen latex kit (Microgen Ltd., Guildford, UK) which permits identification by a slide agglutination reaction in just a few minutes. Cross-reactions with this reagent to *C. bifermentans*, *C. sordellii* and *C. glycolicum* (BRAZIER

1990) have been documented, but in combination with the tests described above, it is a useful additional discriminatory test. A few species of clostridia are commonly mis-identified as *C. difficile*. These include *C. innocuum*, *C. glycolicum* and *C. bifermentans/sordellii*. FEDORKO and WILLIAMS (1997) recently evaluated production of proline-aminopeptidase by a disc test as a means of rapid recognition of *C. difficile*. In conjunction with typical growth on CCFA and morphological characteristics, they claimed this method was useful.

Some workers consider cytotoxin assays to be less sensitive than culture for the diagnosis of *C. difficile* infection. DELMEE et al. (1992) reported only 71% sensitivity of cytotoxin assay in stools compared to isolation of a toxigenic strain of *C. difficile*, and others have published similar findings (CHANG et al. 1979). The value of isolating *C. difficile* from stool specimens as a means of diagnosis irrespective of method, has been the topic of much debate in the literature. The points for and against culture were neatly summarised in an exchange of letters published in the European Journal of Clinical Microbiology, to which the reader is referred (RILEY et al. 1995a). There are valid arguments put by both sides, and, as in most issues, cost factors enter in to the equation. The optimum laboratory investigations should include the ability to perform cultural investigations when the need arises, and a compromise procedure is to store toxin positive stools, frozen for up to 1 month. If an outbreak becomes apparent, cultural investigations can be performed retrospectively either locally or in a specialist laboratory to yield isolates for epidemiological and other investigations. Two alternative methods are faecal lactoferrin assay and methylene blue staining of faeces followed by microscopy for leukocytes, which were compared by YONG et al. (1984). Both methods had limited degrees of sensitivity and specificity which precludes their use as routine diagnostic tests.

4 Demonstration of *C. difficile* Toxins in Stools

4.1 Cytotoxin (B) Assay

Although both toxins A and B are cytotoxic, the use of tissue culture for the detection of *C. difficile* toxins in stools has become synonymous with the detection of toxin B. There are a number of reasons for this, one being that tissue culture was used to detect toxic effects produced by *C. difficile* before it was known that toxin A existed, and another being that toxin B is a more potent cytotoxin than toxin A for most cell lines. The toxin-induced changes to cell morphology may differ depending on the cell line. The cytopathic effect (CPE) is a consequence of disruption of the cell cytoskeleton and generally results in loss of cell–cell contact and cell rounding in many cell lines. Some cells are reputedly more sensitive than others. THELESTAM and BRONNEGARD (1980) compared the sensitivity of five cell lines according to the highest titre of crude toxin that produced a CPE. In order of decreasing sensitivity they found MRC-5 (human lung fibroblasts) > intestinal fibroblasts > Chinese hamster ovary > mouse adrenal > mouse neuroblastoma. CHANG et al. (1979), however, found only modest differences in seven cell lines using a toxic stool filtrate, and no difference at all with crude toxin. In practice, assay for faecal cytotoxins of *C. difficile* can be performed using almost any cell line commonly used in clinical microbiology laboratories, such as Vero, HEp2, HeLa etc. In the United States, fibroblast or Chinese hamster ovary cell lines are often used whereas in Europe, Vero cell lines are commonly used and these are considered by many as the most sensitive. The detection of cytotoxin in faeces by tissue culture is regarded as the "gold standard" test to which other diagnostic tests are compared in evaluations of sensitivity and specificity. The actual cell lines used in these evaluations are important, however, and often overlooked. For example, comparison of an alternative test with cytotoxicity using Vero cells may make the alternative method appear less sensitive than would a comparison with a less sensitive cell line. The standard procedure for a cytotoxin assay is to make a roughly 1:10 suspension of the faecal sample in phosphate buffered saline (PBS), centrifuge it to remove large debris and pass the supernatant through a 0.2µm or 0.45µm membrane filter. The centrifugation step can be omitted with the larger pore filter. This filtrate is then inoculated onto the cell monolayer, which is examined after overnight incubation and again after 48h. Generally, 100µl of filtrate is added to cells in tissue culture roll-tubes and 10µl to cell monolayers in microtitre trays. To confirm its specificity, any CPE should ideally be neutralizable with antitoxin to either *C. difficile* or *C. sordellii*, and normal rabbit or normal horse serum used as a negative control. Non-specific CPEs may result from concomitant viruses or physicochemical factors. These are readily differentiated from the effects of *C. difficile* toxins. It must also be remembered that toxins from other enteric pathogens will also cause a CPE, particularly *C. perfringens* enterotoxin (BORRIELLO et al. 1984). Indeed, one of the advantages of Vero cells compared to others, e.g. MRC-5, is that they are sensitive to *C. perfringens* enterotoxin. This is one advantage that the tissue culture bioassay has over enzyme immunoassay (EIA) (see below), i.e. detection of other known toxins or discovery of new ones. The disadvantages are the relative slowness of the test (24–48h) compared to EIA (1–2h) and having to maintain tissue culture cell lines which commonly requires the close co-operation of a virology department, though continuous cell lines are not difficult to maintain in a bacteriology laboratory.

4.2 Counterimmunoelectrophoresis

Counterimmunoelectrophoresis (CIE) is a generic method of detecting soluble antigens or toxins in body fluids that has been applied to the diagnosis of *C. difficile*. The first published evaluation was by RYAN et al. (1980) who examined 50 stool specimens and reported 100% sensitivity compared to cytotoxin, with a specificity of 80%. As CIE was a popular method at that time for rapid diagnosis of several infectious diseases, other reports followed. Although most of the reports using CIE set out to detect toxin in the stool, the method also detected cell surface antigens that cross-reacted with the *C. sordellii* antitoxin used in the CIE test, as demonstrated by strong CIE reactions with non-toxigenic strains of *C. difficile* (WEST and WILKINS 1982). Despite these drawbacks, it was claimed that CIE could form the basis of a reliable test for the presence of *C. difficile* or its toxins in stools.

4.3 Latex Particle Agglutination Assays

The use of latex particles coated with antibody to directly detect bacterial antigens in clinical material was probably a significant factor in the demise of the use of CIE. As a diagnostic method it did not require expensive equipment, and latex particle agglutination (LPA) reactions are much more amenable to routine testing. In 1984, SHARABADI et al. (1984) developed an LPA reagent from purified antitoxin coated with γ-globulin. LPA gave 8.5% false positive and 9.1% false negative reactions compared to faecal cytotoxin assay on 163 specimens. An early commercially produced LPA kit (Marion Scientific, Kansas, USA and Becton Dickinson, Cockeysville, USA) was marketed as detecting toxin A. This test offered an alternative to laboratories deficient in tissue culture facilities. Unfortunately, other workers evaluating its performance discovered that it did not detect toxin A (LYERLY and WILKINS 1986).

4.4 Enzyme Immunoassay for *C. difficile* Toxins in Stools

YOLKEN et al. (1981) developed and evaluated an enzyme immunoassay (EIA) method for the detection of *C. difficile* toxin in faecal samples. The authors reported only three discrepant results in stools from patients who had a history of treatment for *C. difficile* infection, which were EIA positive but cytotoxin negative, concluding that the method did not give false positive results. The promising results of 100% sensitivity and 98.4% specificity encouraged further developments, and a number of commercial diagnostic EIA kits are now available. There are numerous published evaluations of such kits, a selection of which are summarised in Table 1. One of these kits for detection of toxin A (Vidas) is automated, but published evaluations compare unfavourably with other immunoassays (Table 1). Most of the kits are designed to detect toxin A, but there are a few kits that are designed to detect both toxins. Until recently it was not clear if this offered a significant diagnostic advantage since it was believed virtually all toxin-producing strains produced both toxins. However, this is now known not to be the case (KATO et al. 1997; DEPITRE et al. 1993) and that toxin A-negative, toxin B-positive, strains are present in several hospitals in England and Wales, accounting for 3% of the strains referred for typing (BRAZIER et al. 1999). The most common toxin-variable strain, PCR ribotype 17, corresponds to Delmee's Serogroup F, and because of the common practice of testing stool samples for toxin A only, may be significantly underreported.

Table 1. Published percentage specificities and sensitivities of commercially available kits for diagnosis of
C. difficile from faeces compared to cytotoxin assay

Kit	Sample size	Specificity (%)	Sensitivity (%)	Reference
Premier (A)[a]	101	100	71	BORRIELLO et al. (1992)
	170	98	90	YOLKEN et al. (1981)
	228	95	88	WOLFHAGEN et al. (1994)
	410	100	93	MERZ et al. (1994)
Vidas (A)	194	75	63	WHITTIER et al. (1993)
	285	100	65	DeGIROLAMI et al. (1992)
	329	95	71	ARROW et al. (1994)
	945	98.5	73	SHANHOLTZER et al. (1992)
Bartels (A)	329	92	94	ARROW et al. (1994)
	463	95.5	95.1	MATTIA et al. (1993)
	700	96	87	SCHUE et al. (1994)
Tox-A test (A)	329	92	93	ARROW et al. (1994)
	463	93.7	86.6	MATTIA et al. (1993)
	700	95	87	SCHUE et al. (1994)
	355	100	84.6	FORWARD et al. (1994)
	410	93	99	MERZ et al. (1994)
CD-TOX (A)	160	88	92.3	KNAPP et al. (1993)
Oxoid Tox A Test	407	96.9	83.1	BENTLEY et al. (1998)
Cytoclone (A + B)	285	97.8	75.5	DeGIROLAMI et al. (1992)
	945	99.1	83.6	SHANHOLTZER et al. (1992)
	160	93.5	96.2	KNAPP et al. (1993)
	700	99	89	SCHUE et al. (1994)

[a] Parentheses denotes toxin detected by assay

4.5 Other Methods

An alternative method for the rapid detection of toxin A in faeces is an adaptation
of the rationale behind some kits used in biochemical or hormonal investigations
whereby an antigen–antibody reaction generates a colour change. The principle is
based on a monoclonal antibody to toxin A, but the antigen–antibody reaction is
visualised by a precipitate that is formed when a blue latex coated with antibody
forms a complex with toxin. One such example is the Oxoid *C. difficile* Toxin A Test
kit (Unipath, Basingstoke, UK). A multi-centre clinical evaluation reported a
sensitivity of 83% and a specificity of 96.3% for this kit compared to cytotoxin
assay (BENTLEY et al. 1998). When the discrepant results were resolved by positive
culture, these figures improved to 90.4% and 97.8% respectively. Advantages of
this method are that this technique is less technically demanding than an enzyme-
immunoassay, is more amenable to "one-off" testing and the result is available
after 30 min. A disadvantage is the need for a bench-top microcentrifuge which has
proved to be an obstacle in some laboratories, although these may now be supplied
by the manufacturer. The Immunocard (Meridian Diagnostics Inc., Cincinnati,
Ohio, USA) test uses an antibody raised against *C. difficile* glutamate dehydrogenase and consequently is not specific for toxigenic strains. This kit has been
evaluated in several studies. JACOBS et al. (1996) compared it to culture, EIA and
latex agglutination in a study of 231 diarrhoeal stools and found low relative
sensitivity and specificity values compared to isolation of a toxigenic strain of

C. difficile of 60% and 76% respectively. A better performance was reported by
STANECK et al. (1996) in a study of 927 stools when the Immunocard had a sensitivity of 84% compared to the cytotoxin assay. This study did not report on the
specificity of the kit, however, but concluded that the Immunocard was a useful
adjunct for the diagnosis of *C. difficile* infection. It was interesting to note that in
this study, isolation of a toxigenic strain was regarded as the most accurate single
method of laboratory diagnosis. A product recently on the market is Triage Micro
C. difficile panel (Biosite Diagnostics, San Diego, USA) which combines toxin A
and glutamate dehydrogenase in a similar format.

4.6 Molecular Methods

Initially, research centered on the use of the polymerase chain reaction (PCR) to
differentiate toxigenic from non-toxigenic isolates of *C. difficile* (WREN et al. 1990;
KATO et al. 1991; WOLFHAGEN et al. 1993; ALONSO et al. 1997). More recent
research has been aimed mainly at direct amplification in specimens of the genes for
toxin production. A 33 bp oligonucleotide probe designed to detect a sequence of
the toxin B gene was tried by GREEN et al. (1994) using both radiolabelling and
digoxigenin-labelling. Of 196 stools examined, a 96% correlation to stool cytotoxicity was obtained, with a sensitivity and specificity of 83% and 100% respectively.
This compared very closely to a parallel evaluation of a commercial EIA kit which
gave values of 80% and 98% respectively. The approach of GUMERLOCK et al.
(1991) was to use a PCR method with a pair of primers to detect a 270 bp fragment
of the 16S rRNA gene of *C. difficile*, detecting as few as ten *C. difficile* cells in a stool
containing 10^{11} organisms per gram of faeces. It could also discriminate between
C. difficile and related organisms such as *C. bifermentans/sordellii*, but was unable to
differentiate between toxigenic and non-toxigenic strains. KATO et al. (1993) attempted to amplify a fragment of the toxin A gene in stool specimens, but experienced difficulty with PCR inhibitory substances which required prior treatment by
an ion-exchange column. Of 39 stool specimens examined in this study, PCR results
agreed with both culture and cytotoxin results, and the authors suggested PCR
amplification may be an effective method for the laboratory diagnosis of *C. difficile*
infection. However, the necessity of an ionic exchange process for each specimen is a
barrier to adoption of this particular method as a routine. GUMERLOCK et al. (1993)
claimed good specificity and superior sensitivity of direct PCR compared to tissue
culture cytotoxicity using primers which amplified a 399 bp sequence of the toxin B
gene. BOONDEEKHUN et al. (1993) also developed a PCR method to apply directly to
stools, amplifying a 63 bp repetitive sequence of the toxin A gene, and found a 94%
correlation with stools yielding a positive cytotoxin result but also found one "false
positive" discrepant result. ARZESE et al. (1995) used a PCR method to detect toxin
A gene fragments in faeces from patients on a long-term care ward, claiming improved detection of toxigenic strains.
    A different molecular approach was adopted by WOLFHAGEN et al. (1994) who
used a magnetic immuno-PCR assay (MIPA) involving separation of *C. difficile* in

a faecal sample by magnetic coated monoclonal antibody, DNA extraction and PCR amplification with primers aimed at the toxin B gene. Compared to isolation of a cytotoxin producing strain this method had a reported sensitivity of 96.7%, a specificity of 100%, and a negative predictive value of 94.1%. Perhaps the major benefit of direct PCR methods would be in elucidating the epidemiology allowing detection of reservoirs of toxigenic *C. difficile* in very low numbers. One of the problems of a PCR-based approach to diagnosis is that amplification of DNA of toxin gene fragments is indicative of toxin-producing potential, but does not confirm the presence of toxin in the stool. Another barrier to the adoption of PCR diagnosis becoming routine is the change in working practices necessary to prevent contamination, and henceforth, erroneous results.

5 Methods of Typing *C. difficile*

To understand the nosocomial epidemiology of *C. difficile* infection, various typing or fingerprinting methods have been applied. Early methods were necessarily based on phenotypic properties such as antibiograms, and in one of the first documented outbreak investigations BURDON et al. (1982) found a resistance pattern to three antibiotics in isolates from cases on a surgical ward that were distinct from isolates in the rest of the hospital. However, this method is at best only rudimentary and was soon superseded by WUST et al. (1982) who applied plasmid analysis, soluble protein polyacrylamide gel electrophoresis (PAGE), immunoelectrophoresis of extracellular antigens and antibiograms to 16 isolates from related cases of *C. difficile* infection. Using these methods they showed that 12 of the 16 strains were indistinguishable and therefore strongly implied that cross-infection had taken place. A combination of bacteriocin and bacteriophage typing methods was tried by SELL et al. (1983) with limited success. This was followed by POXTON et al.'s (1984) immuno-chemical fingerprinting of EDTA treated cell extracts, TABAQCHALI et al.'s (1984) radio-PAGE of [^{35}S] methionine-labelled proteins, and DELMEE et al.'s (1985) serogrouping using antibodies to cell surface antigens raised in laboratory animals. NAKAMURA et al. (1981) was the first to use serum agglutination as a method by raising three antisera against *C. difficile*. These antisera could differentiate four distinct serovars among 79 isolates from healthy carriers. PAGE and serotyping were essentially combined by HEARD et al. (1986) who used Tabaqchali's nine radio-PAGE types to immunize rabbits to confirm that certain protein bands detected on the gel were immunogenic and strain specific. A multiple typing approach was described by MAHONY et al. (1991) who used bacteriocin, bacteriophage and plasmid analysis; however, this combination of methods still left 16% of 114 isolates untypeable.

These early typing methods were ostensibly developed to investigate local outbreaks, and to attempt to understand the epidemiology of *C. difficile* infection at a local level. Many of these investigations found evidence that a single type was responsible for a number of cases within their hospital, thus demonstrating that *C. difficile* could be a cross-infection problem (WUST et al. 1982; SELL et al. 1983; POXTON et al. 1984; TABAQCHALI et al. 1984). It soon became apparent, however, that whilst these methods were fine for local use, there was a need for typing schemes that could be applied to further our understanding of the epidemiology of *C. difficile* on a wider scale. Tabaqchali's typing scheme using [^{35}S] radio-labelled methionine PAGE (TABAQCHALI et al. 1984), could distinguish nine distinct groups amongst 250 clinical isolates with a 98% typeability rate. Delmee's serogrouping scheme (DELMEE et al. 1985) could assign one of six groups to 99% of 312 strains. Other schemes had less impressive statistics, however, with 31% of strains untypeable by bacteriocin and bacteriophage typing (SELL et al. 1983). Comparisons between typing schemes were sought and MULLIGAN et al. (1988) found good correlation between the types recognised by plasmid profiling, serotyping, PAGE of cell surface antigens and immunoblotting. COSTAS et al. (1994) applied SDS-PAGE of whole cell proteins to 79 isolates in an outbreak investigation and this method yielded a maximum of approximately 40 bands ranging in size from 18 to 100kDa. This investigation showed 60 of the 79 isolates to be indistinguishable. SDS-PAGE of EDTA extracted cell surface antigens was compared to serogrouping by OGUNSOLA et al. (1995) analysing 61 isolates. This method yielded bands of between 30 and 67kDa and split their 79 isolates into 17 groups which generally correlated well with the results of serogrouping, and could in fact differentiate between some members of the same serogroup.

Methods of whole cell analysis, such as pyrolysis mass spectrometry (PMS), have been used as a means of investigating putative *C. difficile* outbreaks (CARTMILL et al. 1992; MAGEE et al. 1993). This method has the advantage that it can cope with a throughput of large numbers of strains and is very discriminatory. Its disadvantages, however, are the initial cost of the equipment and its inability to assign a permanent type to a strain. It is now regarded primarily as a "fingerprinting" method, best used for within-batch comparisons rather than as a categorical typing method. PMS is excellent at ascertaining if a single strain is responsible for an outbreak, however, and CARTMILL et al. (1992) used PMS to investigate two distinct outbreaks in a long-stay geriatric ward and a female medical ward to show that a single strain was responsible for both outbreaks. MAGEE et al. (1993), similarly investigated an outbreak on two elderly care wards where 24 isolates from 16 patients were indistinguishable by PMS. This study also noted good correlation between the antibiogram profile of the outbreak strain and the PMS results.

5.1 Molecular Typing Methods

Molecular typing methods are generally regarded as superior to phenotypic methods in terms of the stability of expression and greater degrees of typeability, and a number of methods have been applied to *C. difficile*. Plasmid profiling has been assessed (MULDROW et al. 1982; CLABOTS et al. 1988) but proved largely unsuccessful due to the sparse distribution of these extra-chromosomal genetic elements within the species. Analysis of chromosomal DNA of *C. difficile* was tried by KUIJPER et al. (1987) who used whole cell DNA restriction endonuclease analysis (REA) using *Hind*III in an investigation which demonstrated cross-infection between two patients in the same room. WREN and TABAQCHALI (1987) assessed the performance of *Hind*III REA digestion against their panel of nine types recognised by [35S] methionine-labelled protein profiling, finding distinct digest patterns for each. PEERBOOMS et al. (1987) examined 33 wild strains by the same method and claimed to be able to distinguish 22 distinct electrophoretic patterns. An alternative restriction enzyme was utilised by DEVLIN et al. (1987) who used *Cfo*I to digest the DNA extracted from 110 clinical isolates. Using this method they identified 38 unique REA patterns and the remaining 72 isolates could be grouped into 12 REA patterns. They also demonstrated that the REA profiles remained stable for nine strains stored in vitro, and in vivo in two chronic cases, over a 4-month period. REA was also used in a study of colonisation and disease progression by JOHNSON et al. (1990) in which all isolates from nine patients with *C. difficile* infection were of two virtually identical REA types called B and B2. O'NEILL et al. (1991) applied REA in a study of cases of relapsing *C. difficile* infection to show that more than half of the apparent relapses were due to infections with a new strain. SAMORE et al. (1994) applied REA to 205 isolates from 106 patients and demonstrated 55 distinct REA patterns indicating a high degree of strain diversity in their population. These reports illustrate that REA is a highly discriminatory and reproducible method; however, it is a technically demanding procedure and laborious, especially for large numbers of isolates. The main problem with REA is that frequent cutting enzymes such as *Hind*III, *Eco*RI, *Cfo*I or *Bam*HI produce complex digestion products (ca. > 50 bands) that make comparisons by eye extremely tedious.

Restriction Fragment Length Polymorphism (RFLP) is an alternative genotypic method that involves initial REA digestion followed by gel electrophoresis and Southern blotting with selected labelled nucleic acid probes to highlight specific restriction site heterogeneity. This method was first applied to *C. difficile* by BOWMAN et al. (1991) and directly compared to REA results in a study of the molecular characterisation of strains from humans, animals and their environments (O'NEILL et al. 1993). This study neatly illustrated the difference in discrimination between these methods, since from a set of 116 isolates, REA could differentiate 34 distinct types compared to only six by RFLP. Thus RFLP, in addition to being a very labour-intensive method offers few advantages as a typing scheme, and REA/RFLP methods have generally been superseded by methods based on the polymerase chain reaction (PCR).

Arbitrarily primed PCR (AP-PCR) is a genotypic method that permits the detection of polymorphisms without prior knowledge of the target nucleotide sequence. The oligonucleotide primers are usually used singly, are quite long (up to 53 bp) and of a non-specific sequence. A closely related method called random amplified polymorphic DNA (RAPD) commonly uses two oligonucleotide primers which are shorter in length (ca. 10 bp) and also of arbitrary sequence. McMILLIN and MULDROW (1992) viewed AP-PCR as a potentially useful method for

C. difficile, while WILKS and TABQCHALI (1994) assessed AP-PCR against their panel of nine radio-PAGE types and also against some wild isolates from an outbreak. Each of the nine radio-PAGE types gave a different AP-PCR banding pattern and all of the outbreak strains had identical banding patterns. KILLGORE and KATO (1994) also used this approach to type 41 isolates from an outbreak of antibiotic-associated diarrhoea (AAD). These isolates had been previously typed by immunoblotting and showed agreement for 33/34 strains, with a further seven isolates which were untypeable by immunoblotting grouped by AP-PCR. CHACHATY et al. (1994) applied this method to 30 unrelated isolates using three experimental primers. These primers gave results for all 30 isolates, and both differentiated 20 distinct types. Unfortunately the results were poorly reproducible and this is a general disadvantage of methods using arbitrary primers. BARBUT et al. (1993) evaluated a RAPD method using two 10 bp primers in an investigation of AAD in AIDS patients. The same PCR profiles were found in 25 isolates from 15 patients suggesting infection with the same strain, and the authors claimed RAPD was easy to perform and an effective way of distinguishing between isolates of *C. difficile*. A study of 20 isolates from a variety of sources including six ATCC type cultures was performed by SILVA et al. (1994) using an arbitrary primer (designated PG-05). They distinguished four PCR types in this set, but there was little epidemiological data associated with such a diverse set with which to evaluate their results. A study in a Polish maternity hospital utilised AP-PCR in conjunction with other methods to investigate neonatal acquisition of *C. difficile* (MARTIROSIAN et al. 1995). From their PCR results they concluded the birth canal was not the major source and that the infants acquired the organism from the hospital environment; a finding that verified those of earlier studies pre-dating molecular typing results.

PCR ribotyping uses specific primers complimentary to sites within the RNA operon and was first applied to *C. difficile* by GURTLER (1993) who targeted the amplification process at the spacer region between the 16S and 23S rRNA regions. This part of the genome has been shown to be very heterogeneous, in contrast to the rRNA genes themselves which are highly conserved. *C. difficile* was shown to possess ten copies of the rRNA genes in its genome which varied not only between strains but also between different copies on the same genome. This study distinguished 14 different PCR ribotypes in a set of 24 isolates, hence the method promised good discrimination, and because the primers were fixed and the targets known, good theoretical reproducibility. It involved a long (18–96h) and complicated denaturing PAGE procedure, however, necessary because the primers produced large amplicons of between 800 and 1200 bp. This method was improved by CARTWRIGHT et al. (1995) who applied it to 102 isolates obtained from 73 symptomatic patients. Using the same primers as GURTLER (1993), their PCR fragments of similar size range could be separated by straightforward agarose gel after 2h electrophoresis. Forty-one different ribotypes were found within the 102 isolates, confirming the good discrimination of this method reported by GURTLER (1993). Furthermore, they demonstrated that the banding patterns were not affected by the quantity of DNA used in the reaction (a problem associated with AP-PCR and RAPD methods), that the PCR ribotype marker was stable and its expression reproducible. This approach was adapted to routine use by O'Neill et al. (1996) who simplified the DNA extraction method, and using modified primers, produced amplicons ranging from 250 to 600 bp followed by straightforward agarose gel electrophoresis. The discriminatory power was compared to Delmee's serogroups and gave different banding patterns for each of the 19 serogroups. It was the most straightforward of the PCR ribotyping methods to perform and gave reproducible results. This method is currently used routinely by the PHLS Anaerobe Reference Unit who provide a typing service for referred isolates, and from over 2000 strains examined, a library consisting of 116 distinct ribotypes has been constructed (Stubbs et al. 1999).

The performance of AP-PCR using two different primers (PG-05 and ARB11) was compared to PCR ribotyping with pulsed field gel electrophoresis (PFGE, see below) used in arbitration, in a study of 39 clinical isolates and ten known serogroups (Collier et al. 1996). Both AP-PCR methods gave comparable results, identifying eight and seven groups respectively, amongst the known serogroups, and 21 and 20 types in the wild strains. PCR ribotyping was deemed the more discriminatory, as it yielded discriminatory profiles for 23 wild types, and when compared to PFGE was found to be in agreement for 80% of strains as opposed to 60% and 44% for the two AP-PCR methods respectively. The authors also highlighted the lack of reproducibility of AP-PCR methods, as discrepancies were noted using the same primers in different laboratories.

PFGE allows the whole genome to be analysed after digestion with rare cutting restriction endonucleases, such as *Sma*I, *Ksp*I, *Sac*II or *Nru*I, which produce up to ten fragment length polymorphisms per strain. Consequently, analysis and comparisons of PFGE gels are relatively simple. PFGE has been applied to many different genera and was used to investigate 22 isolates of *C. difficile* from an outbreak in an elderly care facility and 30 epidemiologically unrelated isolates (Talon et al. 1995). Serogrouping was also performed, and amongst the outbreak strains two epidemic serogroups (C and K) were identified. Two different PFGE patterns were found to belong to serogroup C and three PFGE patterns were found in serogroup K. Of the unrelated isolates, all had different PFGE profiles illustrating the high degree of discrimination of this method. Kato et al. (1996) combined PFGE and immunoblotting to differentiate reinfection from relapse in a 10-year-old child who had suffered four episodes of *C. difficile* infection. There was good correlation between the two methods, and their results indicated that the second episode after a 17-day tapering course of vancomycin represented a relapse with the same strain, but that the third and fourth episodes were infections with different strains. They also typed five separate colonies from each episode to show that infection with multiple strains did not occur. This concurred with the findings of an earlier study (O'Neill et al. 1991). Another study comparing the performance of PFGE with other typing methods was that of van Dijck et al. (1996) who applied it to 56 strains known to belong to serogroup C and compared the performance of RAPD and PCR ribotyping. PFGE was the most discriminatory of the three methods, and a combination of RAPD and PFGE recognised 13 genotypes within serogroup C. PCR ribotyping was thought to have performed poorly due to the very stable nature of the rRNA spacer region alleles within this serotype. Whilst PFGE is very discriminatory, disadvantages include the initial cost of the equipment, the slowness of the electrophoresis procedure and its complexity. Also, many workers have noted that some strains are repeatedly untypeable by PFGE due to degradation of the extracted DNA, the reason for which is unknown. Studies have shown that these PFGE untypeable strains belong to serogroup G (KATO et al. 1996) which corresponds to O'Neill et al.'s PCR ribotype 1.

It has been concluded that PCR ribotyping, although marginally less discriminatory than PFGE, offers considerable advantages in terms of speed and technical ease (COLLIER et al. 1996). Furthermore, it is considered that the inability of PFGE to type some isolates compromises the overall value of this technique, and that PCR ribotyping is more likely than AP-PCR to be the best alternative (COLLIER et al. 1996).

Opinions are divided as to the optimum molecular typing method for *C. difficile*. Whilst any of the above could be successfully applied to local epidemiological investigations, the choice becomes more limited if one considers the methods that could be used for studies on a multi-centre basis which are needed to further our understanding of the epidemiology on national and international scales. Most would agree that PCR-based methods offer many advantages over REA or RFLP methods. Two main schools of thought exist within this field, however, and are divided on the opinion of using random versus specific oligonucleotide primers, with reproducibility being the main issue of concern. Proponents of AP-PCR or RAPD claim the method is reproducible, as within their own institutions they can perform analyses many times over and achieve the same amplicon profiles. Inherent to the method, however, is the essentially random nature of the low temperature annealing process using non-specific primers. Reproducibility of results necessitates that each of the parameters must be rigorously standardised, particularly quantification of the extracted DNA. Even apparently minor changes, such as changing the supplier of *Taq* polymerase or the use of a different thermal cycler machine, may profoundly affect the resulting amplicons. PCR ribotyping which uses specific oligonucleotide primers should, in theory, be more robust, although there is a paucity of information in the literature on multi-centre PCR ribotyping trials using the same primers. PFGE offers an advantage in terms of discrimination and is recognised as a leading typing method for many other bacterial species. However, unless the problem of untypeability of strains is overcome, its applicability for *C. difficile* typing will be somewhat limited.

Clearly, standardisation would greatly enhance surveillance and advance our knowledge of the global epidemiology of *C. difficile* infection. There are attempts to establish some form of standardisation in the nomenclatures ascribed to strains typed by the various methods (BRAZIER et al. 1994). At present the nomenclature applied to describe the various types is uncoordinated and there is a lack of understanding as to how types relate to one another. An international typing study was established to address this problem, involving seven groups of workers from the United Kingdom, Belgium, Australia and the United States. Participants were asked to submit their type strains as delineated by their own methods, which

included: radio-PAGE, immunoblotting, REA, serogrouping, RFLP, PCR ribotyping, and AP-PCR. These were checked, blind coded and together with some
wild-type isolates, 100 strains were distributed to each group. Each group typed the
set by their own method and submitted their results back to the study coordinator.
The preliminary findings of the study were revealing (BRAZIER et al. 1997). Because
of the large number and diverse origins of the set of strains, many of the groups
encountered new strains not previously recognised by their own typing methods.
This suggests that there are more types of *C. difficile* in existence than was previously appreciated by each group acting individually. There was complete correlation between the results of the three typing schemes which are all dependent either
directly or indirectly on cell surface proteins. For example, serogroup D corresponded to Radio-PAGE type Y and to immunoblot type 7^1, whereas serogroup G
corresponded to types B and 4^1. This study also revealed that certain types were
common to each typing method, indicating distribution of the same types in hospitals in the United Kingdom, Belgium, United States and Australia.

6 Surveillance of *C. difficile* Infection

The incidence of *C. difficile* infection in hospitals is not recorded accurately,
therefore the true extent of disease is unknown. The only recorded parameters in
England and Wales are the number of positive laboratory tests for *C. difficile* toxins
in stool samples and isolation of the organism as reported to the Public Health
Laboratory Service (PHLS) Communicable Disease Surveillance Centre (CDSC).
Figures published in the PHLS Communicable Disease Report for 1998 were 17%
up on the same period for 1997, continuing the upward trend of the last 5 years
(Fig. 1). Figure 3 shows the geographical distribution of hospitals in England and
Wales who have requested outbreak investigations and the distribution of the most
common PCR ribotype 1. Relatively little is known about the distribution of the
different strains of *C. difficile* circulating the hospitals in England and Wales but
data are emerging. From a total of over 2000 isolates originating from all sources,
116 distinct PCR ribotypes have been identified. Analysis of 1400 isolates examined

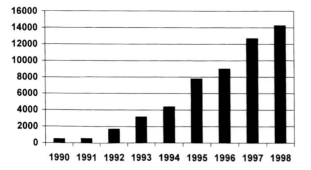

Fig. 1. Laboratory reports of
Clostridium difficile toxin positive stool specimens in England
and Wales between 1990 and
1998

to date, from 40 hospitals, has yielded some interesting statistics. In total, 54 different PCR ribotypes have been identified from hospital patients, but 90% of these are composed of just 16 types, and one particular PCR ribotype, Type 1, accounts for 68% of the total of all hospital patient isolates. The next most common, PCR ribotype 15, accounts for just 5% (Fig. 2). PCR Type 1 appears to be endemic in almost all of the hospitals surveyed and is associated with both acute and prolonged outbreaks. It was PCR Type 1 that was responsible for the most publicised outbreak in the United Kingdom involving 175 patients and 17 deaths in a hospital in the northwest of England (CARTMILL et al. 1994). This outbreak was investigated by three methods, PMS, AP-PCR and PCR ribotyping, which all confirmed that a single outbreak strain was responsible. A study into the relatedness of PCR Type 1 was performed by BRAZIER et al. (1997a) who examined 40 strains from 20 different hospitals by restriction enzyme digestion of PCR products. Eleven different restriction enzymes were used and no differences were found. This indicates a high degree of homogeneity within this PCR ribotype but further work is needed to prove clonality. It was revealed from the International Typing Study (BRAZIER et al. 1997) that Delmee's serogroup G corresponds to PCR Type 1 and this strain is also causing problems in the United States. PCR Type 1 was found to be the same as strain D1 described by SAMORE et al. (1996) (and M.H. Samore, personal communication). They found this was the most common strain isolated from environmental sources, personnel hand carriage and symptomatic patients in an East Coast tertiary referral hospital. This same type has also predominated in a series of 59 isolates from elderly male patients in a hospital in California (M.E. Mulligan, personal communication). The prevalence of this strain in Britain is in contrast to findings of some other European countries. Delmee's group reported that serogroup C was most commonly implicated in outbreaks in Belgium (VAN DIJCK et al. 1996). This serogroup corresponds to PCR Type 12, which accounts for only 2.6% of typed hospital isolates in England and Wales. A multi-centre study of 11 hospitals in France (BARBUT et al. 1996) found serogroups C, D, G and H were the most common strains, with serogroup H predominant (accounting for 21%), and that serogroup C was most often associated with antibiotic treatment and diarrhoea. While many studies have shown that a cluster of cases of *C. difficile* infection was due to a single strain (WUST et al. 1982; POXTON et al. 1984; CARTMILL et al.

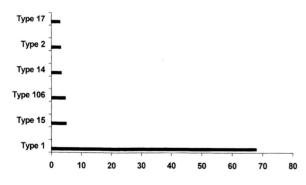

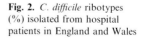

Fig. 2. *C. difficile* ribotypes (%) isolated from hospital patients in England and Wales

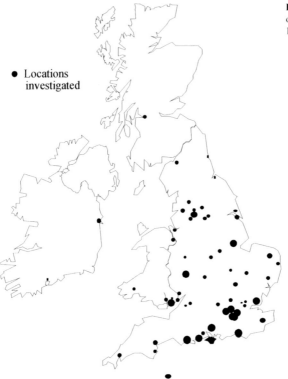

Fig. 3. Geographic distribution of *C. difficile* PCR ribotype 1 in the United Kingdom

● Locations
  investigated

1992; MAGEE et al. 1993; WILKS and TABAQCHALI 1994; KILLGORE and KATO 1994), others have demonstrated that clusters of cases have been due to unrelated strains (SAMORE et al. 1994; BARBUT et al. 1996). These sporadic cases demonstrate that not all cases of *C. difficile* infection are due to cross-infection and most likely represent the diverse strains brought in from the community as explained previously.

The relative virulence to man of the various *C. difficile* types is a topic on which our knowledge is incomplete. The first direct evidence that some types were more virulent than others was produced by BORRIELLO et al. (1987) using the hamster model of disease. There followed a study on the relative virulence of ten *C. difficile* serogroups in hamsters by DELMEE and AVESANI (1990) who reported that types A, C, H and K killed challenged hamsters, while type G and non-toxigenic strains B, D, I and X resulted in faecal colonisation without disease. We now know that non-toxigenic serogroup D strains, which equate to PCR Type 10 and Radio-PAGE type Y, are frequently isolated from neonates and small children. It is also known that serogroup F produces toxin B but not toxin A, and they type as PCR Type 17 by the method of O'NEILL et al. (1996). KATO et al. (1997) looked specifically for the prevalence of these toxin A−/B+ strains in three hospitals in Japan using PCR primers for both toxins. Of 243 isolates, 143 were toxigenic and 47 (33%) were toxin A−/B+. By immunoblotting, all these strains belonged to serogroup F. In the three hospitals these strains accounted for 32%, 13% and 12% of the isolates, although it was not stated if they were isolated from symptomatic patients. Preliminary data from strains received for typing in England and Wales indicate a lower incidence of toxin-variable strains. Toxin A−/B+ strains account for 3% of the total hospital isolates examined; although in one particular hospital they accounted for 10% of the total isolates submitted for typing and overall have been detected in 9/40 hospitals. It is possible that these strains are not being detected because of the common use of diagnostic kits which detect toxin A only, and are far more prevalent than we currently appreciate.

Strains originating from GP patients and controls in England show a different distribution of PCR ribotypes compared to those found in hospital patients. The most predominant strain in a community survey of 370 isolates was PCR Type 10 which is non-toxigenic and accounted for 15.9% of isolates. PCR Type 1 which accounts for 68% of the hospital patient isolates, made up only 7.4% of the 370 community patient isolates. Compared to the overwhelming predominance of one strain in United Kingdom hospitals, the profile of types in the community was far more even, with PCR ribotypes 10, 20 and 14 the most common, accounting for 15.9%, 11.8% and 8.7% respectively. This indicates that certain strains seem to proliferate in hospitals and may even be selected for by local environmental pressures in the hospital ward. An interesting observation on PCR Type 1 is its resistance to erythromycin, a variable characteristic of the species. Another reason could be that hospital strains sporulate better and therefore survive longer in the inanimate environment than less proliferate sporing strains. Alternatively, they may be selected for by increased virulence, thus eliciting more severe symptoms promoting their transmission via contaminated fomites. More studies are needed to fully understand the epidemiology of this organism, including why certain strains appear to have proliferated and remain endemic in many hospitals.

References

Al Saif N, Brazier JS (1996) The distribution of *Clostridium difficile* in the environment of South Wales. J Med Microbiol 45:133–137

Alonso R, Munoz C, Pelaez T, Cercenado E, Rodriquez-Creixems M, Bouza E (1997) Rapid detection of toxigenic *Clostridium difficile* strains by a nested PCR of the toxin B gene. Clin Microbiol Infect 3:145–147

Anonymous (1998) *Clostridium difficile* in England and Wales. Communicable Disease Report. 9, 7, 12 February 1999

Aronsson B, Molby R, Nord CE (1985) Antimicrobial agents and *Clostridium difficile* in acute enteric disease: epidemiological data from Sweden, 1980–1982. J Infect Dis 151:476–481

Arrow SA, Croese L, Bowman RA, Riley TV (1994) Evaluation of three commercial enzyme immunoassay kits for detecting faecal *Clostridium difficile* toxins. J Clin Pathol 47:954–956

Arzese A, Trani G, Riul L, Botta GA (1995) Rapid polymerase chain reaction method for specific detection of toxigenic *Clostridium difficile*. Eur J Clin Microbiol Infect Dis 14:716–719

Aspinall ST, Hutchinson DN (1992) New selective medium for isolating *Clostridium difficile* from faeces. J Clin Pathol 45:812–814

Barbut F, Mario N, Meyohas MC, Binet D, Frottier J, Petit JC (1993) Investigation of a nosocomial outbreak of *Clostridium difficile* – associated diarrhoea among AIDS patients by random amplified polymorphic DNA (RAPD) assay. J Hosp Infect 26:181–189

Barbut F, Corthier G, Charpak Y, Cerf M, Monteil H, Fosse T (1996) Prevalence and pathogenicity of *Clostridium difficile* in hospitalized patients. Arch Intern Med 156:1449–1454

Bartlett JG, Onderdonk AB, Cisneros RL, Kasper DL (1977) Clindamycin associated colitis due to a toxin-producing species of *Clostridium* in hamsters. J Infect Dis 136:701–705

Bartlett JG, Chang TW, Gurwith M, Gorbach SL, Onderdonk AB (1978) Antibiotic associated pseudomembranous colitis due to toxin producing clostridia. N Engl J Med 298:531–534

Bartlett JG (1988) Introduction. In: Rolfe RD, Finegold SM (eds) *Clostridium difficile*: its role in intestinal disease. San Diego Academic Press, San Diego

Bartlett JG (1997) *Clostridium difficile* infection: pathophysiology and diagnosis. Semin Gastrointest Dis 8:12–21

Bartley SL, Dowell VR Jr (1991) Comparison of media for the isolation of *Clostridium difficile* from faecal specimens. Lab Med 22:335–338

Bentley AH, Patel NB, Sidorczuk M, Loy P, Fulcher J, Dexter P, Richards J (1998) Eur J Clin Microbiol Infect Dis 17:788–790

Berry AP, Levett PN (1986) Chronic diarrhoea in dogs associated with *Clostridium difficile* infection. Vet Rec 118:102–103

Blawat F, Chylinski G (1958) Pathogenic clostridia in soil and faeces of domestic animals in the Gdansk region. Bull Inst Marine Med (Gdansk) 9:117–126

Bolton RP, Tait SK, Dear PRF, Lowsowsky MS (1984) Asymptomatic neonatal colonisation by *Clostridium difficile*. Arch Dis Child 59:466–472

Boondeekhun HS, Gurtler V, Odd ML, Wilson VA, Mayall BC (1993) Detection of *Clostridium difficile* enterotoxin gene in clinical specimens by the polymerase chain reaction. J Med Microbiol 38:384–387

Borriello SP, Honour P (1981) Simplified procedure for routine isolation of *Clostridium difficile* from faeces. J Clin Pathol 34:1124–1127

Borriello SP, Honour P, Turner T, Barclay F (1983) Household pets as a potential reservoir for *Clostridium difficile*. J Clin Pathol 36:84–87

Borriello SP, Ketley JM, Mitchell TJ, Barclay FE, Welch AR, Price AB (1987) *Clostridium difficile* – a spectrum of virulence and analysis of putative virulence determinants in the hamster model of antibiotic-associated colitis. J Med Microbiol 24:53–64

Borriello SP, Larson HE, Welch AR, Barclay F, Stringer MF, Bartholomew BA (1984) Enterotoxigenic *Clostridium perfringens*: a possible cause of antibiotic associated diarrhoea. Lancet 1:305–307

Borriello SP, Vale T, Brazier JS, Hyde S, Chippeck E (1992) Evaluation of a commercial enzyme immunoassay kit for the detection of *Clostridium difficile* toxin A. Eur J Clin Microbiol Infect Dis 11:360–363

Bowman RA, Riley TV (1986) Isolation of *Clostridium difficile* from stored specimens and comparative susceptibility of various tissue culture cell lines to cytotoxin. FEMS Microbiol Lett 34:31–35

Bowman RA, Riley TV (1988) Laboratory diagnosis of *Clostridium difficile* associated diarrhoea. Eur J Clin Microbiol Infect Dis 7:476–484

Bowman RA, O'Neill GL, Riley TV (1991) Non-radioactive restriction fragment length polymorphism (RFLP) typing of *Clostridium difficile*. FEMS Microbiol Lett 63:269–272

Brazier JS (1990) Cross reactivity of *Clostridium glycolicum* with the latex particle agglutination reagent for *C. difficile* identification. In: Borriello SP (ed) Clinical and molecular aspects of anaerobes. Wrightson Biomedical Publishing, Petersfield, pp 293–296

Brazier JS (1993) Role of the laboratory in investigations of *Clostridium difficile* diarrhea. Clin Infect Dis 16[Suppl 4]:S228–233

Brazier JS, Delmee M, Tabaqchali S, Hill LR, Mulligan ME, Riley TV (1994) Proposed unified nomenclature for *Clostridium difficile* typing. Lancet 343:157

Brazier JS, Mulligan ME, Delmee M, Tabaqchali S (1997) Preliminary findings of the international typing study on *Clostridium difficile*. Clin Infect Dis 25[Suppl 2]:S199–201

Brazier JS, O'Neill GL, Duerden BI (1997a) Polymerase chain reaction ribotypes of *Clostridium difficile* in hospitals in England and Wales. Reviews in Medical Microbiology 8[Suppl 1]:S55–56

Brazier JS, Stubbs SLJ, Duerden BI (1999) Prevalence of toxin A negative/toxin B positive *Clostridium difficile* strains. J Hosp Infect 42:248–249

Brooks JB, Nunez-Monteil OL, Basta MT, Hierholzer JC (1984) Studies of stools from pseudomembranous colitis rotaviral and other diarrhoeal syndromes by frequency-pulsed electron capture liquid chromatography. J Clin Microbiol 20:549–560

Buchanan AG (1984) Selective enrichment broth culture for detection of *Clostridium difficile* and associated cytotoxin. J Clin Microbiol 20:74–76

Buggy BP, Wilson KH, Fekety R (1983) Comparison of methods for recovery of *Clostridium difficile* from an environmental source. J Clin Microbiol 18:348–352

Burdon DW (1982) *Clostridium difficile*: the epidemiology and prevention of hospital-acquired infection. Infection 10:203–204

Byl B, Jacobs F, Strulens MJ, Thys JP (1996) Extraintestinal *Clostridium difficile* infections. Clin Infect Dis 22:712

Carroll SM, Bowman RA, Riley TV (1983) A selective broth for *lostridium difficile*. Pathology 15:165–167

Cartmill TDI, Orr K, Freeman R, Sisson PR, Lightfoot NF (1992) Nosocomial infection with *Clostridium difficile* investigated by pyrolysis mass spectrometry. J Med Microbiol 37:352–356

Cartmill TDI, Panigrahi H, Worsley MA, McCann DC, Nice CN, Keith E (1994) Management and control of a large outbreak of diarrhoea due to *Clostridium difficile*. J Hosp Infect 27:1–15

Cartwright CP, Stock F, Beekmann SE, Williams EC, Gill VJ (1995) PCR amplification of rRNA intergenic spacer regions as a method for epidemiologic typing of *Clostridium difficile*. J Clin Microbiol 33:184–187

Chachaty EP, Sauliner P, Martin A, Mario N, Andremont A (1994) Comparison of ribotyping, pulsed-field gel electrophoresis and random amplified polymorphic DNA for typing *Clostridium difficile* strains. FEMS Microbiol Lett 122:61–68

Chang TW, Lauermann M, Bartlett JG (1979) Cytotoxic assay in antibiotic-associated colitis. J Infect Dis 140:765–770

Clabots CR, Peterson LR, Gerding DN (1988) Characterization of a noscomial *Clostridium difficile* outbreak by using plasmid profile typing and clindamycin susceptibility testing. J Infect Dis 158:731–736

Clabots CR, Bettin KM, Peterson LR, Gerding DN (1991) Evaluation of cycloserine-cefoxitin fructose agar and cycloserine-cefoxitin fructose broth for recovery of *Clostridium difficile* from environmental sites. J Clin Microbiol 29:2633–2635

Collier MC, Stock F, De Girolami PC, Samore MH, Cartwright CP (1996) Comparison of PCR-based approaches to molecular analysis of *Clostridium difficile*. J Clin Microbiol 34:1153–1157

Costas M, Holmes B, Ganner M, On SL, Hoffman PN, Worsley MA (1994) Identification of outbreak-associated and other strains of *Clostridium difficile* by numerical analysis of SDS-PAGE protein patterns. Epidemiol Infect 113:1–12

Dansinger ML, Johnson S, Jansen PC, Opstad NL, Bettin KM, Gerding DN (1996) Protein-losing enteropathy is associated with *Clostridium difficile* diarrhea but not with asymptomatic colonization: a prospective, case-control study. Clin Infect Dis 22:932–937

De Girolami PC, Hanff PA, Eichelberger K, Longhi L, Teresa H, Pratt J (1992) Multicenter evaluation of a new enzyme immunoassay for detection of *Clostridium difficile* enterotoxin A. J Clin Microbiol 30:1085–1088

Delmee M, Homel M, Wauters G (1985) Serogrouping of *Clostridium difficile* strains by slide agglutination. J Clin Microbiol 21:323–327

Delmee M, Avesani V (1990) Virulence of ten serogroups of *Clostridium difficile* in hamsters. J Med Microbiol 33:85–90

Delmee M, Mackey T, Hamitou A (1992) Evaluation of a new commercial *Clostridium difficile* toxin A enzyme immunoassay using diarrhoeal stools. Eur J Clin Microbiol Infect Dis 11:246–249

Department of Health and Public Health Laboratory Service Joint Working Group (1994) *Clostridium difficile* infection. Prevention and Management. BAPS Health Publication Unit, Heywood Lancashire

Depitre C, Delmee M, Avesani V, L'Haridon R, Roels A, Popoff M, Corthier G (1993) Serogroup F strains of *Clostridium difficile* produce toxin B but not toxin A. J Med Microbiol 38:434–441

Devlin HR, Au W, Foux L, Bradbury WC (1987) Restriction endonuclease analysis of nosocomial isolates of *Clostridium difficile*. J Clin Microbiol 25:2168–2172

Donta ST, Myers MG (1982) *Clostridium difficile* toxin in asymptomatic neonates. J Pediatr 100:431–434

El Mohandes AE, Keiser JF, Refat M, Jackson BJ (1993) Prevalence and toxigenicity of *Clostridium difficile* isolates in fecal microflora of preterm infants in the intensive care nursery. Biol Neonate 63:225–229

Elsden SR, Hilton MG, Waller JM (1976) The end products of the metabolism of aromatic amino acids by clostridia. Arch Microbiol 107:283–288

Fedorko DP, Williams EC (1997) Use of cycloserine-cefoxitin fructose agar and L-proline amino-peptidase (PRO Discs) in the rapid identification of *Clostridium difficile*. J Clin Microbiol 35:1258–1259

Feldman RJ, Kallich M, Wienstein MP (1995) Bacteremia due to *Clostridium difficile*: case report and review of extraintestinal *C. difficile* infections. Clin Infect Dis 20:1560–1562

Forward KR, Dalton MT, Kerr E, Paisley N, Cooper G (1994) Comparison of TechLab *Clostridium difficile* Tox-A enzyme immunoassay and Bartels Prima system toxin A EIA. Diagn Microbiol Infect Dis 20:1–5

Frazier KS, Herron AJ, Hines ME, Gaskin JM, Altman NH (1993) Diagnosis of enteritis and entero-toxemia due to *Clostridium difficile* in captive ostriches (Struthio camelus). J Vet Diagn Invest 5: 623–625

George WL, Sutter VL, Goldstein EJC, Ludwig SL, Finegold SM (1978) Etiology of antimicrobial agent associated colitis. Lancet I:802–803

George WL, Sutter VL, Citron D, Finegold SM (1979) Selective and differential medium for isolation of *Clostridium difficile*. J Clin Microbiol 19:214–219

Gerding DN, Brazier JS (1993) Optimal methods for identifying *Clostridium difficile* infections. Clin Infect Dis 16[Suppl 4]:S439–442

Green GA, Riot B, Monteil H (1994) Evaluation of an oligonucleotide probe and an immunological test for direct detection of toxigenic *Clostridium difficile* in stool samples. Eur J Clin Microbiol Infect Dis 13:576–581

Green RH (1974) The association of viral activation with penicillin toxicity in guinea pigs and hamsters. Yale J Biol Med 3:166–181

Gumerlock PH, Tang YJ, Meyers FJ, Silva J Jr (1991) Use of the polymerase chain reaction for the specific and direct detection of *Clostridium difficile* in human feces. Rev Infect Dis 13:1053–1060

Gumerlock PH, Tang YJ, Weiss JB, Silva J Jr (1993) Specific detection of toxigenic strains of *Clostridium difficile* in stool specimens. J Clin Microbiol 31:507–511

Gurtler V (1993) Typing of *Clostridium difficile* strains by PCR-amplification of variable length 16S-23S rDNA spacer regions. J Gen Microbiol 139:3089–3097

Hafiz S (1974) *Clostridium difficile* and its toxins. (Thesis, PhD) Department of Microbiology, University of Leeds

Hafiz S, McEntegart MG, Morton RS, Waitkins SA (1975) *Clostridium difficile* in the urogenital tract of males and females. Lancet I:420–421

Hafiz S, Oakley CL (1976) *Clostridium difficile*: isolation and characteristics. J Med Microbiol 9:129–136

Hall IC, O'Toole E (1935) Intestinal flora in newborn infants with a description of a new pathogenic anaerobe, *Bacillus difficilis*. Am J Dis Child 49:390–402

Hammarstrom S, Perlmann P, Gustafsson BE, Lagercrantz R (1969) Autoantibodies to colon in germfree rats monocontaminated with *Clostridium difficile*. J Exp Med 129:747–756

Heard SR, Rasburn B, Matthews RC, Tabaqchali S (1986) Immunoblotting to demonstrate antigenic and immunogenic differences among nine standard strains of *Clostridium difficile*. J Clin Microbiol 24:384–387

Hirschorn LR, Trnka Y, Onderdonk AB, Lee ML, Platt R (1994) Epidemiology of community-acquired *Clostridium difficile*-associated diarrhea. J Infect Dis 169:127–133

Holst E, Helin I, Mardh PA (1981) Recovery of *Clostridium difficile* from children. Scandinavian J Infect Dis 13:41–45; Infection with *Clostridium difficile* investigated by pyrolysis mass spectrometry. J Med Microbiol 37:352–356; Investigation of a nosocomial outbreak of *Clostridium difficile* by pyrolysis mass spectrometry. J Med Microbiol 39:345–351

Jacobs J, Rudensky B, Dresner J, Berman A, Sonneblick M, van Dijk Y, Yinnon AM (1996) Comparison of four laboratory tests for diagnosis of *Clostridium difficile*-associated diarrhea. Eur J Clin Microbiol Infect Dis 15:561–566

Johnson S, Clabots CR, Linn FV, Olson MM, Peterson LR, Gerding DN (1990) Nosocomial *Clostridium difficile* colonisation and disease. Lancet 336:97–100

Kato N, Ou CY, Kato H (1991) Identification of toxigenic *Clostridium difficile* by the polymerase chain reaction. J Clin Microbiol 29:33–37

Kato N, Ou CY, Kato H (1993) Detection of toxigenic *Clostridium difficile* in stool specimens by the polymerase chain reaction. J Infect Dis 167:455–458

Kato H, Kato N, Watanabe K, Ueno K, Sakata Y, Fujita K (1996) Relapses or reinfections: analysis of a case of *Clostridium difficile*-associated colitis by two typing systems. Curr Microbiol 33:220–223

Kato H, Kato N, Fukui K, O'Hara A, Watanabe K (1997) High prevalence of toxin A-negative/toxin-B positive *Clostridium difficile* strains among adult inpatients. Clin Microbiol Infect 3[Suppl 2]:S220

Katz DA, Bates DW, Rittenberg E, Onderdonk A, Sands K, Barefoot LA, Snydman D (1997) Predicting *Clostridium difficile* stool cytotoxin results in hospitalized patients with diarrhea. J Gen Intern Med 12:57–62

Kauffman L, Weaver RH (1960) Use of neutral red fluorescence for the identification of colonies of clostridia. J Bacteriol 79:292–294

Killgore GE, Kato H (1994) Use of arbitrary primer PCR to type *Clostridium difficile* and comparison of results with those by immunoblot typing. J Clin Microbiol 32:1591–1593

Kim KH, Fekety R, Batts DH, Brown D, Cudmore M, Silva J Jr (1981) Isolation of *Clostridium difficile* from the environment of contacts of patients with antibiotic-associated colitis. J Infect Dis 143:42–50

Knapp CC, Sandin RL, Hall GS, Ludwig MD, Rutherford I, Washington JA (1993) Comparison of vidas *Clostridium difficile* toxin-A assay and premier *C. difficile* toxin-A assay to cytotoxin-B tissue culture assay for the detection of toxins of *C. difficile*. Diagn Microbiol Infect Dis 17:7–12

Kuijper EJ, Oudbier JH, Stuifbergen WNHM, Jansz A, Zanen HC (1987) Application of whole-cell DNA restriction endonuclease profiles to the epidemiology of *Clostridium difficile* induced diarrhea. J Clin Microbiol 25:751–753

Larson HE, Parry JV, Price AB, Davies DR, Dolby J, Tyrell DA (1977) Undescribed toxin in pseudo-membranous colitis. Br Med J 1:1246–1248

Larson HE, Price AB, Honour P, Borriello SP (1978) *Clostridium difficile* and the aetiology of pseudo-membranous colitis. Lancet I:1063–1065

Larson HE, Barclay FE, Honour P, Hill ID (1982) Epidemiology of *Clostridium difficile* in infants. J Infect Dis 146:727–733

Levett PN (1984) Detection of *Clostridium difficile* in faeces by direct gas liquid chromatography. J Clin Pathol 37:117–119

Levett PN (1985) Effect of antibiotic concentration in a selective medium on the isolation of *Clostridium difficile* from faecal specimens. J Clin Pathol 38:233–234

Levett PN (1986) *Clostridium difficile* in habitats other than the human gastro-intestinal tract. J Infect 12:253–263

Lyerly DM, Wilkins TD (1986) Commercial latex test for *Clostridium difficile* toxin A does not detect toxin A. J Clin Microbiol 23:622–623

Madewell BR, Tang YJ, Jang S, Madigan JE, Hirsh DC, Gumerlock PH (1995) Apparent outbreaks of *Clostridium difficile*-associated diarrhea in horses in a veterinary medical teaching hospital. J Vet Diagn Invest 7:343–346

Magee JT, Brazier JS, Hosein IK, Ribeiro CD, Hill DW, Griffiths A (1993) An investigation of a nosocomial outbreak of *Clostridium difficile* by pyrolysis mass spectrometry. J Med Microbiol 39:345–351

Mahony DE, Clow J, Atkinson L, Vakharia N, Schlech WF (1991) Development and application of a multiple typing system for *Clostridium difficile*. Appl Environ Microbiol 57:1873–1879

Malamou-Ladas H, O'Farrell SO, Nash JQ, Tabaqchali S (1983) Isolation of *Clostridium difficile* from the patients and the environment of hospital wards. J Clin Pathol 36:88–92

Manabe YC, Vintez JM, Moore RD, Merz C, Charache P, Bartlett JG (1995) *Clostridium difficile* colitis: an efficient clinical approach to diagnosis. Ann Intern Med 123:835–840

Martirosian G, Kuipers S, Verburgh H, van Belkum A, Meisel-Mikolajczyk FM (1995) PCR ribotyping and arbitrarily primed PCR for typing strains of *Clostridium difficile* from a Polish maternity hospital. J Clin Microbiol 33:2016–2021

Mattia AR, Doern GV, Clark J, Holden J, Wu L, Ferraro MJ (1993) Comparison of four methods in the diagnosis of *Clostridium difficile* disease. Eur J Clin Microbiol Infect Dis 12:882–886

McBee RH (1960) Intestinal flora of some antarctic birds and mammals. J Bacteriol 79:311–312

McCluskey J, Riley TV, Owen ET, Langlands DR (1982) Reactive arthritis associated with *Clostridium difficile*. Aust N Z J Med 12:535–537

McFarland LV, Surawicz CM, Stamm WE (1990) Risk factors for *Clostridium difficile* carriage and *C. difficile*-associated diarrhea in a cohort of hospitalised patients. J Infect Dis 162:678–684

McMillin DE, Muldrow LL (1992) Typing of toxic strains of *Clostridium difficile* using DNA finger-printings generated with arbitrary polymerase chain reaction primers. FEMS Microbiol Lett 92:5–10

Merz CS, Kramer C, Forman M, Gluck L, Mills K, Senft K (1994) Comparison of four commercially available rapid enzyme immunoassays with cytotoxin assay for detection of *Clostridium difficile* toxin(s) from stool specimens. J Clin Microbiol 32:1142–1147

Muldrow LL, Archibold ER, Nunez-Monteil OL, Sheeny RJ (1982) Survey of the extrachromosomal gene pool of *Clostridium difficile*. J Infect Dis 16:637–640

Mulligan ME, Peterson LR, Kwok RY, Clabots CR, Gerding DN (1988) Immunoblots and plasmid fingerprints compared with serotyping and polyacrylamide gel electrophoresis for typing *Clostridium difficile*. J Clin Microbiol 26:41–46

Nakamura S, Mikawa M, Nakashio S, Takabatake M, Okada I, Yamakawa K (1981) Isolation of *Clostridium difficle* from feces and the antibody in sera of young and elderly adults. Microbiol Immun 25:345–351

Nakamuara S, Serikawa T, Mikawa M, Nakashio S, Yamakawa K, Nishida S (1981) Agglutination, toxigenicity and sorbitol fermentation of *Clostridium difficile*. Microbiol Immunol 25:863–870

O'Neill GL, Beaman MH, Riley TV (1991) Relapse versus reinfection with *Clostridium difficile*. Epidemiol Infect 107:627–635

O'Neill GL, Adams JE, Bowman RA, Riley TV (1993) A molecular characterisation of *Clostridium difficile* isolates from humans, animals and their environments. Epidemiol Infect 111:257–264

O'Neill GL, Ogunsola FT, Brazier JS, Duerden BI (1996) Modification of a PCR ribotyping method for application as a routine typing scheme for *Clostridium difficile*. Anaerobe 2:205–209

Ogunsola FT, Ryley HC, Duerden BI (1995) Sodium dodecyl sulfate – polyacrylamide gel electrophoresis anlaysis of EDTA-extracted cell-surface protein antigens is a simple and reproducible method for typing *Clostridium difficile*. Clin Infect Dis 20[Suppl 2]:S327–330

Peerbooms PGH, Kuijt P, Maclaren DM (1987) Application of chromosomal restriction endonuclease digest analysis for use as a typing method for *Clostridium dificile*. J Clin Pathol 40:771–776

Pepersack F, Labbe M, Nonhoff C, Schoutens E (1983) Use of gas liquid chromatography as a screening test for toxigenic *Clostridium difficile* in diarrhoeal stools. J Clin Pathol 36:1233–1236

Phillips KD, Rogers PA (1981) Rapid detection and presumptive identification of *Clostridium difficile* by *p*-cresol production on a selective medium. J Clin Pathol 34:642–644

Potvliege C, Lable M, Yourassowsky E (1981) Gas liquid chromatography as a screening test for *Clostridium difficile*. Lancet 1:1105

Poxton IR, Byrne MD (1981) Immunochemical analysis of the EDTA-soluble antigens of *Clostridium difficile* and related species. J Gen Microbiol 122:41–46

Poxton IR, Aronsson B, Molby R, Nord CE, Collee JG (1984) Immunochemical fingerprinting of *Clostridium difficile* strains isolated from an outbreak of antibiotic-associated colitis and diarrhoea. J Med Microbiol 17:317–324

Renshaw AA, Stelling JM, Doolittle MH (1996) The lack of value of repeated *Clostridium difficile* cytotoxicity assays. Arch Pathol Lab Med 120:49–52

Riley TV, Adams JE, O'Neill GL, Bowman RA (1991) Gastrointestinal carriage of *Clostridium difficile* in cats and dogs attending veterinary clinics. Epidemiol Infect 107:659–665

Riley TV, Wetherall F, Bowman J, Mogyorosy J, Colledge CL (1991) Diarrhoeal disease due to *Clostridium difficile* in general practice. Pathology 23:346–349

Riley TV (1994) The epidemiology of *Clostridium difficile*-associated diarrhoea. Rev Med Microbiol 5:117–126

Riley TV, Cooper M, Bell B, Golledge CL (1995) Community-acquired *Clostridium difficile* – associated diarrhea. Clin Infect Dis 20[Suppl 2]:S263–265

Riley TV, Bowman RA, Golledge CL (1995a) Usefulness of culture in the diagnosis of *Clostridium difficile* infection. Eur J Clin Microbiol Infect Dis 14:1109–1111

Ryan RW, Kwasnik I, Tilton RC (1980) Rapid detection of *Clostridium difficile* toxin in human feces. J Clin Microbiol 12:776–779

Samore MH, Bettin KM, DeGirolami PC, Clabots CR, Gerding DN, Karchmer AW (1994) Wide diversity of *Clostridium difficile* types at a tertiary referral hospital. J Infect Dis 170:615–621

Samore MH, Vankataraman L, De Girolami PC, Arbeit RD, Karchmer AW (1996) Clinical and molecular epidemiology of sporadic and clustered cases of nosocomial *Clostridium difficile* diarrhea. Am J Med 100:32–40

Schue V, Green GA, Monteil H (1994) Comparison of the Tox-A test with cytotoxicity assay and culture for the detection of *Clostridium difficile*-associated diarrhoea disease. J Med Microbiol 41:316–318

Sell TL, Schaberg DR, Fekety FR (1983) Bacteriophage and bacteriocin typing scheme for *Clostridium difficile*. J Clin Microbiol 17:1148–1152

Shanholtzer CJ, Peterson LR, Olsen MM, Gerding DN (1983) Prospective study of gram stained stool smears in diagnosis of *Clostridium difficile* colitis. J Clin Microbiol 17:906–908

Shanholtzer CJ, Willard KE, Holter JJ, Olson MM, Gerding DN, Peterson LR (1992) Comparison of VIDAS *Clostridium difficile* toxin A immunoassay with *C. difficile* culture and cytotoxin and latex tests. J Clin Microbiol 30:1837–1840

Sharabadi MS, Bryan LE, Gaffney D, Coderre SE, Gordon R, Pai CH (1984) Latex agglutination test for detection of *Clostridium difficile* toxin in stool samples. J Clin Microbiol 20:339–341

Sheretz RJ, Sarubbi MD (1982) The prevalence of *Clostridium difficile* and toxin in a nursery population: a comparison between patients with necrotizing enterocolitis and an asymptomatic group. J Pediatr 100:435–439

Silva J Jr, Yajarayma JT, Gumerlock PT (1994) Genotyping of *Clostridium difficile* isolates. J Infect Dis 169:661–664

Smith LDS, King EO (1962) Occurrence of *Clostridium difficile* in infections of man. J Bacteriol 84:65–67

Spencer RC (1998) Clinical impact and associated costs of *Clostridium difficile*-associated disease. J Antimicrob Chemother 41[Suppl C]:S5–12

Staneck JL, Weckbach LS, Allen SD, Siders JA, Gilligan PH, Coppitt G, Kraft JA, Willism DH (1996) Multicenter evaluation of four methods for *Clostridium difficile* detection: immunocard C. *difficile*, cytotoxin assay, culture and latex agglutination. J Clin Microbiol 34:2718–2721

Stevenson JP (1966) The normal bacterial flora of the alimentary canal of laboratory stocks of the desert locust *Schistocerca gregaria* Forskal. J Invertebr Pathol 8:205–211

Struble AL, Tang YJ, Kass PH, Gumerlock PH, Madewell BR, Silva J Jr (1994) Fecal shedding of *Clostridium difficle* in dogs: a period of prevalence survey in a veterinary medical teaching hospital. J Vet Diagn Invest 6:342–347

Stubbs SLJ, Brazier JS, O'Neill GL, Duerden BI (1999) PCR targeted to the 16S-23S rRNA gene intergenic spacer region of *Clostridium difficile* and construction of a library consisting of 116 different PCR ribotypes. J Clin Microbiol 37:461–463

Tabaqchali S, Holland D, O'Farrell S, Silman R (1984) Typing scheme for *Clostridium difficile*: its application in clinical and epidemiological studies. Lancet I:935–938

Talon D, Bailly P, Delmee M, Thouverez M, Mulin B, Iehl-Robert M (1995) Use of pulsed-field gel electrophoresis for investigation of an outbreak of *Clostridium difficile* infection among geriatric patients. Eur J Microbiol Infect Dis 14:987–993

Tedesco FJ, Barton RW, Alpers DH (1974) Clindamycin-associated colitis. Ann Intern Med 81:429–433

Thelestam M, Bronnegard M (1980) Interaction of cytopathogenic toxin from *Clostridium difficile* with the cells in tissue culture. Scand J Infect Dis Suppl 22:16–29

van Dijck P, Avesani V, Delmee M (1996) Genotyping of outbreak-related and sporadic isolates of *Clostridium difficile* belonging to serogroup C. J Clin Microbiol 34:3049–3055

West S, Wilkins TD (1982) Problems associated with counterimmunoelectrophoresis assays for detecting *Clostridium difficile* toxin. J Clin Microbiol 15:347–349

Whittier S, Shapiro DS, Kelly WF, Walden TP, Wait KJ, McMillon LT (1993) Evaluation of four commercially available enzyme immunoassays for laboratory diagnosis of *Clostridium difficile*-associated diseases. J Clin Microbiol 31:2861–2865

Wilcox MH, Smyth ETM (1998) UK survey of the incidence and impact of *Clostridium difficile* infection 1993–1996 (abstract). In: Programme of the Third Federation of Infection Societies Conference, Manchester, p 20

Wilks M, Tabaqchali S (1994) Typing of *Clostridium difficile* by polymerase chain reaction with an arbitrary primer. J Hosp Infect 28:231–234

Willey SH, Bartlett JG (1979) Cultures for *Clostridium difficile* in stools containing a cytotoxin neutralised by *Clostridium sordellii* antitoxin. J Clin Microbiol 10:880–884

Wilson KH, Silva J, Fekety FR (1982) Fluorescent-antibody test for detection of *Clostridium difficile* in stool specimens. J Clin Microbiol 16:464–468

Wolfhagen MJHM, Fluit AC, Jansze M, Rademaker CMA, Verhoef J (1993) Detection of toxigenic *Clostridium difficile* in fecal samples by colony blot hybridization. Eur J Clin Microbiol Infect Dis 12:463–466

Wolfhagen MJ, Fluit AC, Torensma R, Poppelier MJ, Verhoef J (1994) Rapid detection of toxigenic *Clostridium difficile* in fecal samples by magnetic PCR assay. J Clin Microbiol 32:1629–1633

Wren BW, Tabaqchali S (1987) Restriction endonuclease DNA analysis of *Clostridium difficile*. J Clin Microbiol 25:2402–2404

Wren BW, Clayton CL, Tabaqchali S (1990) Rapid identification of toxigenic *Clostridium difficile* by polymerase chain reaction. Lancet 335:423

Wust J, Sullivan NM, Hardegger U, Wilkins TD (1982) Investigation of an outbreak of antibiotic-associated colitis by various typing methods. J Clin Microbiol 16:1096–1101

Yolken RH, Whitcomb LS, Marien G (1981) Enzyme immunoassay for the detection of *Clostridium difficile* antigen. J Infect Dis 144:378

Yong WH, Mattia AR, Ferarro MJ (1994) Comparison of fecal lactoferrin latex agglutination assay and methylene blue microscopy for detection of fecal leukocytes in *Clostridium difficile* associated disease. J Clin Microbiol 32:1360–1361

Genetics of *Clostridium difficile* Toxins

J.S. MONCRIEF and T.D. WILKINS

1 Introduction

Up until the time it was implicated as the cause of pseudomembranous colitis (PMC), *C. difficile* was an almost unknown species of bacteria. The organism was first isolated from healthy newborn infants in 1935 by Hall and O'Toole, who named it *Bacillus difficilis* after the apparent difficulty they encountered in its isolation (HALL and O'TOOLE 1935). The organism produced a toxic culture filtrate which was lethal to animals upon injection. Perhaps due to the fact that the organism did not appear to cause disease, few studies followed its initial discovery. Over 40 years later, in 1977, *C. difficile* was implicated as the cause of a lethal colitis that resulted from treatment with antibiotics. The discovery resulted from the

Virginia Polytechnic Institute and State University, West Campus Drive, Fralin Biotechnology Centre, Blacksburg, VA 24061-0346, USA and TechLab Inc., VPI Research Park, 1861 Pratt Drive, Blacksburg, VA 24060-6364, USA

finding that antisera to *C. sordellii* neutralized toxic activity found in fecal filtrates from patients with antibiotic-associated colitis. Curiously, however, *C. sordellii* could not be isolated from the patients. *C. difficile*, on the other hand, had been isolated previously from many patients, but had been ignored since it was "non-pathogenic". Further investigation showed that toxic activity in culture filtrates from *C. difficile* was neutralized by *C. sordellii* antisera. Thus the fortuitous cross-neutralizing activity of *C. sordellii* antisera led to the discovery of *C. difficile* as the cause of antibiotic-associated colitis.

Continued research revealed that *C. difficile* actually produces two toxins. Both toxins, which came to be known as toxins A and B, are unusually large proteins and share a number of similar features. We now know that *C. sordellii* produces two very similar toxins, thus explaining the cross-neutralizing activity of antisera observed in early studies on antibiotic-associated colitis. The *C. difficile* and *C. sordellii* toxins, along with the alpha toxin of *C. novyi* comprise a group called the large clostridial toxins. They share extensive sequence identity and have a common mechanism of action involving glucosylation of the low molecular weight GTP-binding proteins that control a number of cellular functions. This chapter will review the genetics of the *C. difficile* toxins and also briefly discuss the other large clostridial toxins as well as a newly discovered putative large glucosyltransferase toxin from hemorrhagic *Escherichia coli*.

2 Molecular and Biological Properties of *Clostridium difficile* Toxins

Toxins A and B have been purified and their molecular and biological properties have been extensively characterized (Table 1). They both cause similar rounding of tissue culture cells. Toxin B is active against every cell line which has been tested. Toxin A, though far less cytotoxic on most cell lines when compared with

Table 1. Molecular and biological properties of *Clostridium difficile* toxins

Property	Toxin A	Toxin B
Mr	308,000	270,000
Pi	5.6	4.2
Cytotoxicity[a]	10ng	1pg
Lethal dose[b]	50ng	50ng
Enterotoxicity[c]	1μg	_[d]
Hemagglutination[e]	+	−
Glucosyltrasferase	+	+

[a] Minimum dose causing > 50% rounding of CHO-KI cells.
[b] Minimum lethal dose for intraperitoneal challenge of mice.
[c] Minimum dose causing fluid secretion in rabbit ileal loop assay.
[d] Toxin B from variant strain 8864 is enterotoxic.
[e] Agglutination of rabbit erythrocytes at 4°C.

toxin B, is a potent cytotoxin for epithelial cells of intestinal origin such as HT-29 cells (Tucker et al. 1990). These cells have a high density of carbohydrate receptors for toxin A. Rounding of intoxicated cells is accompanied by disassembly of the actin microfilaments. Microtubules and intermediate filaments are only secondarily affected (Fiorentini et al. 1990; Mitchell et al. 1987; Ottlinger and Lin 1988; Thelestam and Bronnegard et al. 1980). Although the microfilament cytoskeleton is preferentially affected, actin itself is not the target. Rearrangement of actin and the actin binding proteins vinculin and talin occurs. Both toxins are glucosyltransferases that covalently modify the Rho proteins (Dillon et al. 1995; Just et al. 1994, 1995a,b,c). These low molecular weight GTP-binding proteins function in organization of the cytoskeleton, explaining the toxin effect on the cytoskeleton.

Toxin A is an extremely potent enterotoxin with microgram amounts causing fluid secretion in animal loops (Libby et al. 1982; Lima et al. 1988; Lyerly et al. 1982; Lyerly and Wilkins 1995). On a molar basis it is as active as cholera toxin. Its mechanism of action, however, is quite different. Unlike cholera toxin, toxin A causes extensive damage to the epithelial lining of the intestine. The villus tips of the epithelium are initially disrupted, followed by damage to the brush border membrane. The mucosa eventually becomes denuded, accompanied by extensive infiltration with neutrophils. Massive inflammation results and undoubtedly plays an important role in pathogenesis. The fluid response is partly due to damage to the intestinal epithelium. The cytotoxic activity of toxin A also results in disruption of the tight junctions, which likely plays an important role in mechanism of toxin A enterotoxicity (Hecht et al. 1988; Hipppensteil et al. 1997). Toxin A also has been shown to elicit the production of various biological response modifiers such as cytokines and neurokinins, which appear to play an important role in pathogenesis (Castaglioulo et al. 1994, 1997, 1998; Manyth et al. 1996a,b; Pothoulakis et al. 1994, 1998).

Toxin B does not cause damage or a fluid response when injected alone in intestinal loops (Lima et al. 1988; Lyerly et al. 1982; Lyerly and Wilkins 1995). This is probably due to an inability to bind to a receptor on the intestinal brush border membrane cells under normal physiological conditions. Toxin A binds to specific carbohydrate receptors on the surface of intestinal cells and initiates damage to the intestine. Toxin B then contributes to the extensive damage during the course of disease, once it gains access to the underlying tissue. This role for toxin B in pathogenesis is supported by experiments on the effects of the toxins in animal models (Lyerly et al. 1985). In these studies, toxin A caused death when given intragastrically to hamsters, whereas toxin B had no effect by this route. On the other hand, when toxin B was given along with sublethal amounts of toxin A, the animals died with no observable intestinal pathology. Moreover, if the intestine is physically disrupted, toxin B gains access to the systemic circulation and causes death. There remains controversy about the role of the toxins in human disease. It has been reported that toxin B is more potent than toxin A to the human colonic epithelium in vitro (Riegler et al. 1995). The intestinal tissue used in these experiments may have been damaged during extraction allowing toxin B access to the epithelial cells. Therefore, whether or not this is relevant to what happens in the intact human intestine remains to be determined.

Toxins A and B have similar lethal activity when injected intraperitoneally with both toxins having a minimum lethal dose of about 50ng in mice (EHRICH 1982; LYERLY et al. 1982; LYERLY and WILKINS 1995). The mechanism by which the toxins cause death when administered systemically is unknown. In mice injected intraperitoneally there is some evidence of liver damage. Arnon et al. examined the effects of the toxins in rhesus monkeys and noted that the animals apparently died from cessation of breathing (ARNON et al. 1984). Death is not accompanied by the paralysis characteristic of the botulinum and tetanus neurotoxins.

Toxin A is more resistant than toxin B to proteases such as trypsin and chymo-trypsin and to extremes of pH (LYERLY et al. 1986). Both toxins can be protected by reducing agents and are inactivated by oxidizing agents. Toxin B has a high negative surface charge at neutral pH, whereas toxin A has very little surface charge.

3 Cloning of the *C. difficile* Toxin A and B Genes

Before the genes of toxins A and B were cloned, there was considerable controversy about the size of *Clostridium difficile* toxins. Many had trouble believing the toxins could be so large and assumed that like most other toxins they must be composed of multiple subunits. Cloning and sequencing of the toxin A and B genes laid the controversy to rest and revealed a great deal about the structure and similar nature of the toxins.

In the mid-1980s, a number of researchers began work on cloning the *C. difficile* toxin genes. Cloning and expression of a 0.3kb fragment of the toxin A gene in a lambda gt11 library was reported in 1987 (MULDROW et al. 1987). Wren et al. also reported lambda gt11 clones containing a toxin A gene fragment (WREN et al. 1989). The clone expressed a 235kDa protein that caused elongation of Chinese hamster ovary cells. In both cases, however, the clones were unstable. Eichel-Streiber et al. reported cloning of overlapping fragments of the toxin A gene (EICHEL-STREIBER et al. 1988). We reported cloning of the carbohydrate binding portion of toxin A in 1987 (PRICE et al. 1987). The cloned 4.7kb PstI restriction fragment expressed a protein that agglutinated rabbit erythrocytes and reacted with affinity-purified antibody to toxin A. The recombinant fragment, however, did not have cytotoxic activity. The 4.7kb toxin A gene fragment was used as the initial probe for chromosomal walking to clone flanking regions of the *C. difficile* VPI 10463 chromosome (DOVE et al. 1990; JOHNSON et al. 1990). As it turned out, the toxin A and B genes were located in close proximity and this work led to the cloning and sequencing of both toxin genes.

Sequencing of the toxin A gene by Dove et al. showed that the structural gene is 8130 nucleotides and encodes a 308kDa protein which is 2710 amino acids in length (DOVE et al. 1990). This confirmed that toxin A was as large as had been predicted by SDS-PAGE analysis. At that time, it made toxin A the largest known single polypeptide bacterial toxin. The entire toxin A gene was reconstructed from cloned fragments of the toxin A gene and expressed in *E. coli*. Lysates of recombinant *E. coli* expressing the toxin A gene contained the cytotoxic activity, enterotoxic activity, and lethal properties associated with toxin A (PHELPS et al. 1991).

Chromosomal walking experiments used for cloning the toxin A gene led directly to cloning of the toxin B gene (JOHNSON et al. 1990). The 3′ end of the toxin B gene turned out to be located 1350bp upstream of the toxin A initiation codon. A clone that contained the 5′-end of the toxin A gene and a small open reading frame (ORF) was found to also contain 1.2kb of DNA which, when subcloned, expressed a nontoxic peptide that reacted with toxin B antibodies. The rest of the toxin B gene was isolated on a 6.8kb fragment that overlapped this gene fragment. The full-length toxin B gene was reconstructed from the two overlapping fragments and expressed in *E. coli*. The recombinant protein reacted with toxin B antisera and was cytotoxic and lethal. The sequence of the toxin B gene revealed a structural gene of 7098bp with a deduced amino acid sequence of 2366 amino acids (M_r 270,000) in length (BARROSO et al. 1990). The gene sequence confirmed that toxin B is also an unusually large protein expressed as a single polypeptide. The toxin A and B genes have also been cloned from VPI 10463 by Eichel-Streiber et al. (EICHEL-STREIBER et al. 1988, 1992a; EICHEL-STREIBER 1995).

4 Sequence Identity and Conserved Features of the Toxins

A comparative sequence analysis of the toxin A and B genes confirmed their relatedness at the structural level (EICHEL-STREIBER et al. 1992a; EICHEL-STREIBER 1995). Alignment of the amino acid sequences showed extensive sequence identity. The toxins share 49% identity and are 63% similar when conservative substitutions are considered. The sequence identity is striking in some areas. The extensive sequence identity and close proximity of the toxin genes suggest that the genes are the result of gene duplication (EICHEL-STREIBER 1995).

4.1 N-terminal Glucosyltransferase Domain

Protein toxins typically are composed of an N-terminal enzymatic domain and a C-terminal receptor binding domain. It was assumed therefore that the enzymatic activity of toxins A and B would be found in the N-terminal portion of the toxins. This meant that there was a region of approximately 200kDa upstream of the toxins' binding domains in which the enzymatic activity might be contained due to the unusually large size of the toxins.

A small hydrophobic region is located approximately in the center of this 200kDa region. We therefore suspected that the enzymatic activity might be located in the N-terminal third of the toxins (Fig. 1). The first evidence for this came from our studies on a deletion mutant of toxin B (BARROSO et al. 1994). This mutant toxin B was deleted for a large part of the middle region of the toxins, but contained the N-terminal third of toxin B fused in-frame with the repeating units. Although activity of the mutant toxin was dramatically reduced, it retained some cytotoxicity. This indicated that the enzymatic activity was indeed contained in the N-terminal third of the holotoxin. Confirmation of this came from a detailed analysis of in vitro glucosyltransferase activity of various recombinant fragments from within this region.

The region containing enzymatic activity was found to reside entirely within the first 546 amino acids (63kDa) of toxin B (HOFMANN et al. 1997). The 63kDa fragment had as much glucosyltransferase activity as the holotoxin and caused cell rounding when microinjected into NIH-3T3 cells. These findings were confirmed and extended in additional studies (WAGENKNECHT-WIESNER et al. 1997). The enzymatic activity of toxin A is likely contained within this relatively small region, although this has not been confirmed. Toxins A and B contain a DXD motif that contains aspartic acid residues believed to participate in catalytic activity. The DXD motif is part of a larger glucosyltransferase sequence motif determined to be the active site cleft. In the closely related toxin LT from *C. sordellii* the two aspartic

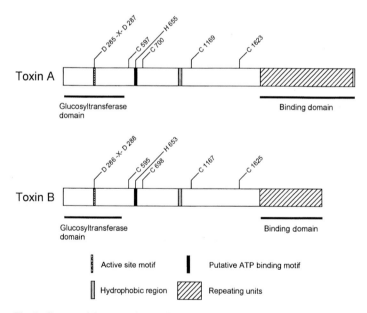

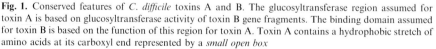

Fig. 1. Conserved features of *C. difficile* toxins A and B. The glucosyltransferase region assumed for toxin A is based on glucosyltransferase activity of toxin B gene fragments. The binding domain assumed for toxin B is based on the function of this region for toxin A. Toxin A contains a hydrophobic stretch of amino acids at its carboxyl end represented by a *small open box*

acid residues (D286 and D288) have been shown to be essential for enzymatic activity (BUSCH et al. 1998).

4.2 Repeating Units

Toxins A and B enter the cell primarily by receptor-mediated endocytosis (FLORIN et al. 1983; HENRIQUES et al. 1987). A large series of repeating units is contained at the carboxyl end of both toxins. In the case of toxin A, the repeating units bind specific cell-surface carbohydrate receptors (POTHOULAKIS et al. 1996; PRICE et al. 1987). The repeating units of toxin A form a multivalent lectin that agglutinates rabbit erythrocytes. Its specificity for rabbit erythrocytes, which contain the carbohydrate Galα1-3Galβ1-4GlcNAc, led to the discovery of the carbohydrate portion of toxin A receptors (KRIVAN et al. 1986). Toxin A binds to rabbit erythrocytes at 4°C but not at 37°C. This is probably due to the low density of receptors on the red cell surface. The affinity of the repeating units for a single carbohydrate receptor appears to be low at 37°C. Effective binding at physiological temperatures, therefore, may require a high density of receptors. As a result, toxin A may preferentially target specific cell lines, such as intestinal epithelial cells and leukocytes which contain a high density of receptors.

Toxin A also binds to human carbohydrate antigens I, X, and Y, all of which have a type 2 core structure that is conformationally similar to Galα1-3Galβ1-4GlcNAc (SMITH et al. 1997; TUCKER and WILKINS 1991). These antigens are all expressed by human intestinal epithelial cells and may function as receptors for toxin A. The X antigen is also present in high amounts on the surface of human neutrophils and other leukocytes, indicating that toxin A may also target these cells. Receptors for toxin B have not been identified. The potent activity of toxin B on a broad range of cells suggests that the receptor(s) is ubiquitous.

The repeating units of toxin A comprise nearly one-third of the carboxyl-terminus. This region contains a series of 38 contiguous repeating units composed of 831 amino acids. Based on length and low levels of sequence similarity, the repeating units were grouped into classes I and II (DOVE et al. 1990). Toxin A contains 7 class I repeating units (each 30 amino acids in length) and 31 class II repeats (each containing 20 or 21 amino acids). The class II repeats were subdivided into four groups, A, B, C, and D. The repeating units share homology with the carbohydrate binding region of streptococcal glucosyltransferases (EICHEL-STREIBER et al. 1990, 1992b). The streptococcal glucosyltransferases and repeating units of toxins A and B are rich in aromatic amino acids and with rare exceptions each unit contains the sequence YF. The YF sequence was used to align the repeating units for classification into subgroups based on additional sequence similarity (DOVE et al. 1990; EICHEL-STREIBER et al. 1990). It has been proposed that the repeats have a modular design in which the aromatic amino acids function in primary protein–carbohydrate interaction (EICHEL-STREIBER et al. 1992b). The affinity for binding is amplified by the repetition of the sequences. Binding specificity is determined by a second, as yet undefined, characteristic of the arrangement and sequence of the repetitive structure.

In this manner the proteins have evolved an ability to bind different carbohydrate structures based on a similar fundamental unit.

4.3 Additional Conserved Features

Toxins A and B share several additional structural features within the approximately 150kDa region between the glucosyltransferase region and the repeating units. These include four conserved cysteines, a potential nucleotide binding site, and a central hydrophobic region (Fig. 1). Additional clusters with exceptional sequence identity are also present within the region between the hydrophobic area and the beginning of the repeating units (EICHEL-STREIBER 1995). The function of the four conserved cysteines is not known. Reducing agents do not affect the toxins, suggesting that the cysteines are not involved in formation of disulfide bridges critical to toxin structure. Instead, the cysteine residues may function as free sulfhydryls involved in ligand binding. Alternatively, the conserved cysteines may be strategically located near processing points essential for toxin activation inside the cell. Some other bacterial toxins have centrally located hydrophobic regions which are thought to function as membrane-spanning regions necessary for toxin transport. Whether or not the hydrophobic center of toxins A and B has a similar function remains to be determined.

We developed a series of deletion and site-directed mutants of the toxin B gene which altered some of these conserved features and studied the effect on cytotoxicity (BARROSO et al. 1994; MONCRIEF et al. 1997b). To investigate the role of the repeating units and fourth cysteine, regions of varying length were removed from the carboxyl terminal end by exonuclease deletions of the 3' end of the toxin B gene. Removal of the repeating units of toxin B resulted in a tenfold reduction in activity. Surprisingly, however, the mutant toxin still had significant activity, since the recombinant lysate had a cytotoxicity titer of 10^5 compared with 10^6 for the unaltered toxin. Removal of a further 125 amino acids decreased the cytotoxicity another tenfold; however, a significant amount of activity (10^4) still remained. In contrast, deletion of an additional 141 amino acid region, that contained the fourth conserved cysteine, resulted in complete inactivation of the toxin. This suggested that this region and the conserved cysteine are critical to toxin function. Substitution of the fourth cysteine (Cys-1625) with serine resulted in a tenfold decrease in activity supporting this idea.

Other site-directed mutants were developed for the region at the N-terminal third of toxin B which contains the putative nucleotide binding site. When serine was substituted for the second conserved cysteine (Cys-698), located just downstream of the proposed nucleotide binding site, a tenfold decrease in activity was observed. Substitution of a histidine residue (His-653) in the putative nucleotide binding site eliminated more than 99% of cytotoxic activity, despite the fact that this region is not required for glucosyltransferase activity. Taken together these results illustrate the importance of the large central domain of the toxins located between the glucosyltransferase domain and the repeating units.

5 *C. difficile* Toxigenic Element

It has recently emerged that genes encoding bacterial virulence factors are often contained in clusters which are acquired by horizontal gene transfer. These genetic "blocks" or "cassettes" are referred to as pathogenicity islands (GROISMAN and OCHMANN 1996; HACKER et al. 1997). In 1995, Hammond et al. described the boundaries of a pathogenicity island containing the toxin A and B genes, which they termed the toxigenic element (HAMMOND and JOHNSON 1995). The toxigenic element (Fig. 2) of *C. difficile* VPI 10463 is 19.6kb and contains three other small ORFs, termed *txe1* (also *txeR*), *txe2*, and *txe3* in addition to the toxin genes. Later, Hundsberger et al. also described the element in VPI 10463 and referred to it as the pathogenicity locus (PaLoc), with the genes termed *tcdA-E* (HUNDSBERGER et al. 1997). Two of the smaller ORFs, *txeR* and *txe2*, are transcribed in the same direction as the toxin genes (HAMMOND et al. 1996; HUNDSBERGER et al. 1997). Transcript levels of both toxins and these two ORFs peak during stationary phase, suggesting they are coordinately regulated. This is supported by the fact that both toxin gene promoters are positively regulated by *txeR* (see below). The protein encoded by *txe2* contains a potential lipoprotein signal sequence at its N-terminus; its function is unknown. The additional ORF, *txe3*, is transcribed in the opposite direction. Its mRNA transcript peaks during exponential growth and is not present during stationary phase (HUNDSBERGER et al. 1997). It was proposed that it might play a role in negative regulation of the toxin genes although this remains to be determined.

Most pathogenicity islands are comprised of a large region of DNA and encode many virulence-related proteins. Due to its small size, the toxigenic element may be more properly termed a pathogenicity islet. Nontoxigenic strains of *C. difficile* contain a 127 bp sequence at the locus where the toxigenic element is found. This sequence contains inverted repeats with predicted secondary structure and may represent a target for insertion of the toxigenic element. Similarly, a pathogenicity islet is found in enterotoxigenic *Bacteroides fragilis*. Reminiscent of the *C. difficile* toxigenic element, the pathogenicity islet of enterotoxigenic *B. fragilis* contains two very similar genes next to each other. The genes encode a proteolytic enterotoxin termed fragilysin and another very similar metalloprotease (MONCRIEF et al. 1998). Furthermore, nontoxigenic strains of *B. fragilis* contain a 17 bp sequence at this locus which may serve as a target for integration of the fragilysin pathogenicity islet. Thus, two widely divergent anaerobes appear to have acquired enterotoxins by site-specific integration of a pathogenicity islet.

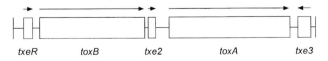

Fig. 2. Toxigenic element of *C. difficile* VPI 10463. *Arrows* indicate the direction of transcription

6 Regulation of *C. difficile* Toxins

It has been known for years that *C. difficile* produces far more toxin when grown under conditions that slow the growth of the organism. Nutrient limitation, such as that which occurs during growth in dialysis sacs, leads to very high levels of toxins A and B. Limitation of biotin in defined media increases the production of *C. difficile* toxins (YAMAKAWA et al. 1996). On the other hand, rapid growth in rich media essentially shuts down toxin production. Toxin expression is also repressed by glucose (DUPRAY and SONENSHEIN 1998).

The promoters of the toxin A and B genes have unusually long distances between the transcriptional start site and the ATG start codons: 169 nucleotides for toxin A and 239 nucleotides for toxin B (EICHEL-STREIBER et al. 1995). Similar spacing is found in the promoter regions of tetanus and botulinum toxin genes as well as the ultraviolet light-inducible promoter P1 of *C. perfringens*. The promoters share some sequence identity with the *C. perfringens* UV-inducible promoter P1 and have been verified as the functional promoters of toxin A and B genes by transcriptional analysis (DUPRAY and SONENSHEIN 1998; HAMMOND et al. 1996; HUNDSBERGER et al. 1997; SONG and FAUST 1998).

The *txeR* gene, located immediately upstream of the toxin B gene, encodes a small (22kDa) basic protein which is rich in lysine residues at its C-terminus. It contains a helix-turn-helix motif within the lysine-rich region which has sequence identity to DNA binding proteins. Additionally, TxeR shares 22% identity with UviA, the putative regulator of the UV-inducible promoter P1. These charac-teristics suggested to us that it may play a role in regulation of toxin gene expression. To determine whether *txeR* regulates expression from the toxin A and B promoters, we fused a DNA fragment encoding the toxin A repeating units (ARU) to each of the promoters and analyzed their expression in *Escherichia coli* (MONCRIEF et al. 1997a). Expression of the reporter gene, with or without the presence of *txeR*, was measured using enzyme-linked immunosorbent assay. When *txeR* was supplied in *trans*, expression from the toxin B promoter-ARU increased over 800-fold. Similarly expression from the toxin A promoter-ARU fusion increased over 500-fold in the presence of *txeR*. In conclusion, *txeR* en-codes a *trans*-acting regulatory protein that positively regulates expression from both toxin promoters.

Based on detailed analysis of toxin gene transcripts, Hammond et al. suggested that the toxins are transcribed as part of a very large (17.5kb) poly-cistronic mRNA (HAMMOND et al. 1996). Transcripts approximately the size of the toxin genes, which were detected at much higher levels, were proposed to result from posttranslational processing. Our studies with *txeR* in *E. coli*, how-ever, suggest that the majority of toxin gene transcription takes place from the individual toxin gene promoters and that high level expression during stationary phase growth is dependent on the *txeR* gene product. The large transcript ob-served may be the result of incomplete transcriptional termination during high level expression.

TxeR also has significant homology to small ORFs located near the tetanus and botulinum neurotoxin genes. These proteins, TetR and BotR, also contain a helix-turn-helix DNA binding motif. They are 29% identical to TxeR and share about 50% identity with each other. Marvaud et al. recently showed that TetR and BotR positively regulate the expression of neurotoxins in *C. tetani* and *C. botulinum*, respectively (MARVAUD et al. 1998a,b). Interestingly, they also showed that BotR increased expression of tetanus toxin when introduced into *C. tetani*. This suggests a conserved mechanism of regulation of the neurotoxin genes of *C. tetani* and *C. botulinum*. Like the *C. difficile* toxins, the botulinum and tetanus neurotoxins are produced in large amounts when grown in dialysis sac cultures, suggesting they are also regulated by nutrient limitation (STERNE and WENTZEL 1950).

TxeR also contains some similarity to particular sigma factors known as the extracellular function (ECF) family (A.L. Sonenshein, personal communication). In Gram-positive organisms, these sigma factors act as general stress response sigma factors (MISSIAKIS and RAINA 1998). TxeR therefore, may not function as a typical response regulator by acting as a transcriptional activator. Rather, TxeR may be a sigma factor that turns on toxin production in response to stress (e.g., limited nutrients). Expression of TxeR itself may be controlled by upstream events. In turn, high level transcription from the toxin gene promoters is initiated when TxeR becomes available. In some cases ECF sigma factors are modulated by a membrane protein which has anti-sigma activity. These membrane proteins are thought to serve as a sensor and signalling molecule which allows an adaptive response to specific environmental conditions. Nothing is known, however, about any other genes or proteins which control the expression of toxins A and B.

7 Variant *C. difficile* Toxin Genes

Toxigenic strains of *C. difficile* almost invariably produce both toxins. In 1991, however, a strain (8864) was identified that produced only toxin B (TORRES 1991). Despite not producing toxin A, 8864 is virulent in hamsters. This may be due to the fact that it produces a variant toxin that is more potent than toxin B from VPI 10463. At the sequence level the toxin from 8864 is more similar to *C. sordellii* LT (SOEHN et al. 1998). It also causes fluid secretion in rabbit intestinal loops (BORRIELLO et al. 1992; LYERLY et al. 1992). These findings suggested that variant strains of *C. difficile* might cause disease.

Initially two different types of toxin A-negative/toxin B-positive strains were identified (DEPITRE et al. 1993; RUPKIN et al. 1997; SOEHN et al. 1998). Strain 8864 has a large deletion (5.6kb) in the toxin A gene. Additionally, it has a DNA insert of about 1.1kb between *txe2* and the beginning of the toxin A gene. The other variant type was that of the serotype F strains (e.g., strain 1470). They have been isolated from infants but do not appear to cause disease in humans or animal models. PCR analysis indicated that they are deleted for only a small portion (approximately 1.7kb) of the toxin A gene within the repeating units. This portion of the repeats contains the epitopes for the toxin A monoclonal antibody (PCG-4) used in toxin A ELISAs, thus explaining why these strains are negative in immunoassays for toxin A (Frey and Wilkins 1992; Lyerly et al. 1983). This region is also responsible for toxin A binding to intestinal cells, offering a reason for the lack of biologically active toxin A.

Rupkin et al. recently published an extensive study in which they characterized over 200 strains of *Clostridium difficile* for restriction length polymorphisms (RFLPs) and deletions in the toxin genes (Rupkin et al. 1998). When compared with reference strain VPI 10463, differences were found in 21% of the strains. Variant strains could be divided into ten groups (toxinotypes I to X). Groups I, II, VI, VII, and VIII all contained deletions within the repeating units of toxin A. Only group VIII, however, was not detected by immunoassay for toxin A, indicating the remaining portion of the repeats in the other strains contained at least one epitope for toxin A monoclonal antibody PCG-4. Therefore, these strains would not be detected as variant strains with current assays used in clinical studies. Strains from group VIII, which did not react in toxin A ELISA, all turned out to be serotype F strains similar to those previously described. Other variant strains described in this study had variations in RFLPs, most of which were found in the toxin B gene. Groups IV through X contained a variation in the promoter region upstream of the regulatory gene *txeR* in addition to changes in RFLPs or deletions found elsewhere.

More recently, evidence has accumulated that toxin A-negative/toxin B-positive strains can cause disease in humans. In a hospital outbreak in Canada toxin A-/toxin B-positive strains were associated with clinical symptoms that ranged from mild diarrhea to fulminant PMC (Alfa et al. 1999; Al-Barrak et al. 1999). Variant strains isolated from symptomatic patients in this and an additional study failed to react in toxin A ELISAs based on monoclonal antibody PCG-4 (Moncrief et al. 1999). However, they did react in a new generation commercial ELISA which detects both toxin A and B. The strains in these studies were deleted for approximately 1.7kb of a region within the repeating units of toxin A, similar to that seen for serotype F strains which also do not react in toxin A ELISAs. In another study in Japan, strains which failed to react in toxin A ELISA were isolated from both symptomatic and asymptomatic patients (Kato et al. 1998). All of these toxin A-negative/toxin B-positive strains were deleted for approximately 0.5kb within the repeating units of toxin A. Additionally, a *C. difficile* strain producing only toxin A has been identified from a patient with AAD (Cohen et al. 1998). In light of these findings, it is important to determine the frequency of antibiotic-associated colitis due to variant strains of *C. difficile*. If these strains are more common than previously thought it raises important issues to address regarding diagnosis as well as treatment and prevention strategies for *C. difficile* infections.

8 Genetic Manipulation of *C. difficile*

The analysis of *C. difficile* virulence factors, including the toxins, has been hampered by the lack of a system for genetic transformation. In this regard, Mullany et al. have reported some success in gene transfer using conjugative transposons (MULLANY et al. 1990, 1991, 1994). Site-specific integration into the *C. difficile* chromosome was accomplished using pCI195, a pBR322-based vector containing a 4.2kb region of the conjugative transposon Tn*919*. Conjugative transfer required a *Bacillus subtilis* strain that contained the related transposon Tn*916*ΔE as an intermediate host. The transposon::plasmid structure was transferable from *B. subtilis* to *C. difficile* by filter mating at a frequency of 10^{-8}. They further demonstrated that pCI195 could be used for gene cloning into *C. difficile* by transfer of a 1.1kb fragment of the toxin B gene into a nontoxigenic strain of *C. difficile*. The transposon::plasmid structure entered the genome at a specific site, as occurs in *B. subtilis*. However, the system is limited to the introduction of small segments of DNA as part of a conjugative transposon. The large size of the toxin genes therefore prohibits their study using this system of conjugal gene transfer. Direct transformation of *C. difficile* with plasmid DNA has not been demonstrated. Cryptic plasmids have been found in some *C. difficile* strains and may provide a basis for developing functional cloning vectors for *C. difficile*.

9 Large Clostridial Toxins of *C. sordellii* and *C. novyi*

C. sordellii and *C. novyi* also produce very large cytotoxins. These toxins include alpha toxin (M_r 250,000) of *C. novyi* and hemorrhagic toxin (HT, M_r 300,000) and lethal toxin (LT, M_r 260,000) of *C. sordellii*. Based on their similarity to the *C. difficile* toxins, BETTE et al. suggested these toxins form a distinct group of closely related toxins (BETTE et al. 1989, 1991). Research within the last few years has established that they have extensive sequence identity and share a common mechanism of action. Accordingly, the glucosyltransferase toxins of *C. difficile*, *C. novyi*, and *C. sordellii* are referred to as the large clostridial toxins (EICHEL-STREIBER 1996).

   C. sordellii toxin HT is very similar to *C. difficile* toxin A (MARTINEZ and WILKINS 1988, 1992). The toxins cross-react immunologically and antisera to each toxin neutralizes the other. HT is cytotoxic, enterotoxic, and lethal at minimal levels similar to those for toxin A. The HT gene has not been sequenced and therefore the extent of homology to toxin A is unknown.

   C. sordellii toxin LT is more closely related to *C. difficile* toxin B (POPOFF 1986; MARTINEZ and WILKINS 1992). The LT amino acid sequence of LT shares 76% identity with toxin B and 47% with toxin A (GREEN et al. 1995). Like toxin B, LT is not an effective enterotoxin. Antisera to LT neutralizes toxin B but not toxin A. Based on biological properties, LT appears to have diverged more from toxin B than HT has from toxin A. For example, LT is about 1000 times less cytotoxic, yet 10 times more lethal than toxin B. LT also has a broader spectrum of glucosylation targets. In addition to Rho, LT glucosylates members of the Ras family of small GTPases (Genth et al. 1996; Hofmann et al. 1996; Just et al. 1996; Popoff et al. 1996). Variants of toxin B isolated from some of the atypical *C. difficile* strains appear to be more closely related to LT than to toxin B from VPI 10463 (Soehn et al. 1998). These variants of toxin B also glucosylate the Ras proteins. The genes of two of these variant toxin Bs have been sequenced. The sequence of the two variants was nearly identical in the glucosyltransferase domain, whereas significant variance was found in this region when they were compared to VPI 10463 toxin B. The majority of amino acid differences were found in the region just downstream of the putative active site. This region has been implicated in Ras recognition by LT and may explain the different activities found in the variant *C. difficile* toxin B (Hofmann et al. 1998).

C. *novyi* alpha toxin gene is carried by a bacteriophage and has been cloned and sequenced (Eklund et al. 1974, 1976; Hofmann et al. 1995). It has 48% identity to both toxin A and B. It also contains similar structural features to toxins A and B. It is a potent cytotoxin and is about ten times as lethal as the *C. difficile* toxins (Ball et al. 1993). Although alpha toxin modifies the same Rho subfamily members as the *C. difficile* toxins, it uses UDP-GlcNAc rather than UDP-Glc as a substrate (Selzer et al. 1996).

10 Putative Large Glucosyltransferase Toxin from Enterohemorrhagic *Escherichia coli*

Complete sequencing of the large 92kb virulence plasmid of *E. coli* O157:H7 recently revealed that enterohemorrhagic *E. coli* carries the gene for an unusually large putative cytotoxin (Burland et al. 1998). The ORF encodes a 3169 amino acid protein of approximately 365kDa. If it is a cytotoxin, it will replace toxin A as the largest known single polypeptide bacterial toxin. The protein sequence contains a glucosyltransferase active site motif (DXD) which is also found in the large clostridial toxins, the OCH1 family of yeast glucosyltransferases, and a putative glucosyltransferase from *Chlamydia trachomatis*. This DXD motif, or a corresponding DXH motif, is common to many glucosyltransferases (Hagen et al. 1999). In addition to the active site motif, the large *E. coli* protein has a hydrophobic region near its center, which like that of the large clostridial toxins contains a predicted membrane-spanning helix.

11 Concluding Remarks

Genetic studies of the *C. difficile* toxins have revealed a lot about their structure and function, however, a great deal remains to be learned. It is intriguing that such large toxins were conserved as a single polypeptide. Approximately the final third of each toxin appears to function in binding. The glucosyltransferase activity necessary for cytotoxicity is contained within the first 63kDa. This leaves a region of approximately 140kDa for which the function is unknown. Mutagenesis studies indicate the vital importance of this region to cytotoxicity. The large middle region likely functions in entry and intracellular trafficking of the toxin. Further studies will continue to shed light on the functions of different regions of the toxins.

The large clostridial toxins covalently modify the Rho and Rac family of low molecular weight GTP-binding proteins that regulate actin microfilaments. It is now appreciated that these GTPases are central players in the regulation of many cellular functions including cell division, motility, intercellular communication, and apoptosis, among others (HALL 1998; MAKAY and HALL 1998). The fact that the targets of these toxins play such a vital role in so many cellular functions makes the large clostridial toxins very useful as tools for studying cell biology. Other clostridial enzymes modify the Rho/Rac family of proteins by ADP-ribosylation (AKTORIES 1997; SEHR et al. 1998). These enzymes, however, do not contain receptor binding domains and are therefore less useful than the large toxins produced by *C. difficile*, *C. sordellii*, and *C. novyi* as molecular tools. Furthermore, although the mechanism of action of the toxins is very similar, each toxin has subtle differences in its effects on cells and the proteins it targets (CHAVES-OLARTE et al. 1997; CIESLA and BOBAK et al. 1998). Moreover, there appear to be a number of variant toxins whose effects on cells can be studied. By dissecting the biochemistry of these toxins' action and effect on cells we should learn a great deal about cell physiology.

References

Aktories K (1997) Rho proteins: targets for bacterial toxins. Trends Microbiol 5:282–288

Al-Barrak A, Embil J, Dyck B, Olekson K, Alfa M, Kabani A (1999) An outbreak of toxin A negative, toxin B positive *Clostridium difficile*-associated diarrhea in a Canadian tertiary-care hospital. Canada Communicable Dis Report 25:65–69

Alfa M, Lyerly D, Neville L, Moncrief S, Al-Barrak A, Kabani A, Dyck B, Olekson K, Embil J (1999) Outbreak of toxin A(−), toxin B(+) *Clostridium difficile*-associated diarrhea in a Canadian tertiary care hospital. Am Soc Microbiol Ann Meet Abstr, 99th.

Arnon SS, Mills DC, Day PA, Henrickson RV, Sullivan NM, Wilkins TD (1984) Rapid death of infant rhesus monkeys injected with *Clostridium difficile* toxins A and B: physiologic and pathologic basis. J Pediatr 104:34–40

Ball DW, Van Tassel RL, Roberts MD, Hahn PE, Lyerly DM, Wilkins TD (1993) Purification and characterization of alpha-toxin produced by *Clostridium novyi* type A. Infect Immun 61:2912–2918

Barroso LA, Wang SZ, Phelps CJ, Johnson JL, Wilkins TD (1990) Nucleotide sequence of the *Clostridium difficile* toxin B gene. Nucleic Acids Res 18:4004

Barroso LA, Moncrief JS, Lyerly DM, Wilkins TD (1994) Mutagenesis of the *Clostridium difficile* toxin B gene and effect on cytotoxic activity. Microbial Pathog 16:297–303

Bette P, Frevert J, Mauler F, Suttorp N, Habermann E (1989) Pharmacological and biochemical studies of the cytotoxicity of *Clostridium novyi* type A alpha toxin. Infect Immun 57:2507–2513

Bette P, Oksche A, Mauler F, Eichel-Streiber Cv, Popoff MR, Habermann E (1991) A comparative biochemical, pharmacological, and immunological study of *Clostridium novyi* alpha-toxin, *C. difficile* toxin B and *C. sordellii* lethal toxin. Toxicon 29:877–887

Borriello SP, Wren BW, Hyde S, Seddon SV, Sibbons P, Krishna MM, Tabaqchali S, Manek S, Price AB (1992) Molecular, immunological, and biological characterization of a toxin A-negative, toxin B-positive strain of *Clostridium difficile*. Infect Immun 60:4192–4199

Burland V, Shao Y, Perna NT, Plunkett G, Sofia HJ, Blattner FR (1998) The complete DNA sequence and analysis of the large virulence plasmid of *Escherichia coli* O157:H7. Nucleic Acids Res 26:4196–4204

Busch C, Hofmann F, Selzer J, Munro S, Jeckel D, Aktories K (1998) A common motif of eukaryotic glycosyltransferases is essential for the enzyme activity of large clostridial toxins. J Biol Chem 273:19566–19572

Castagliuolo I, LaMont JT, Letourneau R, Kelly C, O'Keane JC, Jaffer A, Theoharides TC, Pothoulakis C (1994) Neuronal involvement in the intestinal effects of *Clostridium difficile* toxin A and *Vibrio cholerae* enterotoxin in rat ileum. Gastroenterology 107:657–665

Castagliuolo I, Keates AC, Oiu B, Kelly CP, Nikulasson S, Leeman SE, Pothoulakis C (1997) Increased substance P responses in dorsal root ganglia and intestinal macrophages during *Clostridium difficile* toxin A enteritis in rats. Proc Natl Acad Sci USA 94:4788–4793

Castagliuolo I, Riegler M, Pasha A, Nikulasson, Lu B, Gerard C, Gerard NP, Pothoulakis C (1998) Neurokinin-1 (NK-1) receptor is required in *Clostridium difficile*-induced enteritis. J Clin Invest 101:1547–1550

Chaves-Olarte E, Weidmann M, Eichel-Streiber C, Thelestam M (1997) Toxins A and B from *Clostridium difficile* differ with respect to enzymatic potencies, cellular substrate specificities, and surface binding to cultured cells. J Clin Invest 100:1734–1741

Ciesla WP, Bobak DA (1998) *Clostridium difficile* toxins A and B are cation-dependent UDP-Glucose hydrolases with differing catalytic activities. J Biol Chem 273:16021–1602

Cohen SH, Tang YJ, Hansen B, Silva J Jr (1998) Isolation of a toxin B-deficient mutant strain of *Clostridium difficile* in a case of recurrent *C. difficile*-associated diarrhea. Clin Infect Dis 26:410–412

Depitre C, Delmee M, Avesani V, L'Haridon R, Roels A, Popoff M, Corthier G (1993) Serogroup F strains of *Clostridium difficile* produce toxin B but not toxin A. J Med Microbiol 38:434–441

Dillon ST, Rubin EJ, Yakubovich M, Potholakis C, LaMont JT, Feig LA, Gilbert RJ (1995) Involvement of Ras-Related Rho proteins in the mechanism of action of *Clostridium difficile* toxin A and toxin B. Infect Immun 63:1421–1426

Dove CH, Wang S-Z, Price SB, Phelps CJ, Lyerly DM, Wilkins TD, Johnson JL (1990) Molecular characterization of the *Clostridium difficile* toxin A gene. Infect Immun 58:480–488

Dupray B, Sonenshein AL (1998) Regulated transcription of *Clostridium difficile* toxin genes. Molecular Microbiology 27:107–120

Ehrich M (1982) Biochemical and pathological effects of *Clostridium difficile* toxins in mice. Toxicon 20:983–989

Eichel-Streiber CV, Suckau D, Wachter M, Hadding U (1988) Cloning and characterization of overlapping DNA fragments of the toxin A gene of *Clostridium difficile*. J Gen Microbiol 135:55–64

Eichel-Streiber CV, Sauerborn M (1990) *Clostridium difficile* toxin A carries a C-terminal repetitive structure homologous to the carbohydrate binding region of streptococcal glycosyl-transferases. Gene 96:107–113

Eichel-Streiber CV, Laufenberg-Feldmann R, Sartingen S, Schilze J, Sauerborn M (1992a) Comparative sequence analysis of the *Clostridium difficile* toxins A and B. Mol Gen Genet 233:260–268

Eichel-Streiber CV, Sauerborn M, Kuramitsu HK (1992b) Evidence for a modular structure of the homologous repetitive C-terminal carbohydrate-binding sites of *Clostridium difficile* toxins and *Streptococcus mutans* glycosyltransferases. J Bacteriol 174:6707–6710

Eichel-Streiber CV (1995) Molecular biology of the *Clostridium difficile* toxins. In: Sebald M (ed) Genetics and molecular biology of the anaerobic bacteria. Springer-Verlag, New York pp 264–289

Eichel-Streiber CV, Boquet P, Sauerborn M (1996) Large clostridial cytotoxins – a family of glucosyl-transferases modifying small GTP-binding proteins. Trends Microbiol 4:375–382

Eklund MW, Poysky FT, Meyers JA, Pelroy GA (1974) Interspecies conversion of *Clostridium botulinum* type C to *Clostridium novyi* type A by bacteriophage. Science 186:456–458

Eklund MW, Poysky FT, Peterson ME, Meyers JA (1976) Relationship of bacteriophages to alpha toxin production in *Clostridium novyi* types A and B. Infect Immun 14:793–803

Fiorentini C, Malorni W, Paradisi S, Giuliano M, Mastrantonio P, Donelli G (1990) Interaction of *Clostridium difficile* toxin A with cultured cells: cytoskeletal changes and nuclear polarization. Infect Immun 58:2329–2336

Florin I, Thelestam M (1983) Internalization of *Clostridium difficile* cytotoxin into cultured human lung fibroblasts. Biochim Biophys Acta 763:383–392

Frey S, Wilkins TD (1992) Localization of two epitopes recognized by monoclonal antibody PCG-4 on *Clostridium difficile* toxin A. Infect Immun 60:2488–2492

Genth H, Hofmann F, Selzer J, Rex G, Aktories K, Just I (1996) Difference in protein substrate specificity between hemorrhagic toxin and lethal toxin from *Clostridium sordellii*. Biochem Biophys Res Commun 229:370–374

Green GA, Schue V, Monteil (1995) Cloning and characterization of the cytotoxin L-encoding gene of *Clostridium sordellii*: homology with *Clostridium difficile* toxin B. Gene 161:57–61

Groisman EA, Ochman H (1996) Pathogenicity islands: bacterial evolution in quantum leaps. Cell 87:791–794

Hacker J, Blum-Oehler G, Muhldorfer, Tschape (1997) Pathogenicity islands of virulent bacteria: structure, function and impact on microbial evolution. Mol Microbiol 23:1089–1097

Hagen FK, Hazes B, Raffo R, de Sa D, Tabak LA (1999) Structure and function of the UDP-N-acetyl-D-galactosamine: polypeptide N-acetylgalactosaminyltransferase. Essential residues lie in a predicted active site cleft resembling a lactose repressor fold. J Biol Chem 274:6797–6803

Hall J, O'Toole E (1935) Intestinal flora in newborn infants with description of a new pathogenic organism, *Bacillus difficilis*. Am J Dis Child 135:390–402

Hall A (1998) Rho GTPases and the actin cytoskeleton. Science 279:509–514

Hammond GA, Johnson JL (1995) The toxigenic element of *Clostridium difficile* strain 10463. Microb Pathog 19:203–213

Hammond GA, Lyerly DM, Johnson JL (1996) Transcriptional analysis of the toxigenic element of *Clostridium difficile*. Microb Pathog 22:143–154

Hecht G, Pothoulakis C, LaMont JT, Madara JL (1988) *Clostridium difficile* toxin A perturbs cytoskeletal structure and junction permeability in cultured human epithelial cells. J Clin Invest 82:1516–1524

Henriques B, Florin I, Thelestam M (1987) Cellular internalization of *Clostridium difficile* toxin A. Microb Pathog 2:455–463

Hippensteil S, Tannert-Otto S, Vollrath N, Krull M, Just I, Aktories K, von Eichel-Streiber C, Suttrop N (1997) Glucosylation of the small GTP-binding Rho proteins disrupts endothelial barrier function. Am J Physiol 272:L38–L43

Hofmann F, Habermann E, von Eichel-Streiber C (1995) Sequencing and analysis of the gene encoding the alpha-toxin of *Clostridium novyi* proves its homology to toxins A and B of *Clostridium difficile*. Mol Gen Genet 247:670–679

Hofmann F, Rex G, Aktories K, Just I (1996) The ras-related protein Ral is monoglucosylated by *Clostridium sordellii* lethal toxin. Biochem Biophys Res Commun 227:77–81

Hofmann F, Busch C, Prepens U, Just I, Aktories K (1997) Localization of the glucosyltransferase activity of *Clostridium difficile* toxin B to the N-terminal part of the holotoxin. J Biol Chem 272:11074–11078

Hofmann F, Busch C, Aktories K (1998) Chimeric clostridial cytotoxins: identification of the N-terminal region involved in protein substrate recognition. Infect Immun 66:1076–1081

Hundsberger T, Braun V, Weidmann M, Leukel P, Sauerborn M, Eichel-Streiber CV (1997) Transcription analysis of the genes *tcdA-E* of the pathogenicity locus of *Clostridium difficile*. Eur J Biochem 244:735–742

Johnson JL, Phelps CJ, Barroso LA, Roberts MD, Lyerly DM, Wilkins TD (1990) Cloning and expression of the toxin B gene of *Clostridium difficile*. Curr Microbiol 20:397–401

Just I, Fritz G, Aktories K, Giry M, Popoff MR, Boquet P, Hegenbarth S, Eichel-Streiber CV (1994) *Clostridium difficile* toxin B acts on the GTP-binding protein Rho. J Biol Chem 269:10706–10712

Just I, Selzer J, Eichel-Streiber Cv, Aktories K (1995a) The low molecular mass GTP-binding protein Rho is affected by toxin A from *Clostridium difficile*. J Clin Invest 95:1026–1031

Just I, Selzer J, Wilm M, Eichel Streiber Cv, Mann M, Aktories K (1995b) Glucosylation of Rho proteins by *Clostridium difficile* toxin B. Nature 375:500–503

Just I, Wilm M, Selzer J, Rex G, Eichel Streiber Cv, Mann M, Aktories K (1995c) The enterotoxin from *Clostridium difficile* (ToxA) monoglucosylates the Rho proteins. J Biol Chem 270:13932–13939

Just I, Selzer J, Hofmann F, Green GA, Aktories K (1996) Inactivation of Ras by *Clostridium sordellii* lethal toxin-catalyzed glucosylation. J Biol Chem 271:10149–10153

Kato H, Kato N, Wantabe K, Iwai N, Nakamura H, Yamamoto T, Suzuki K, Kim S-M, Chong Y, Wasito EB (1998) Identification of toxin A-negative, toxin B-positive *Clostridium difficile* by PCR. J Clin Microbiol 36:2178–2182

Krivan HC, Clark GF, Smith DF,Wilkins TD (1986) Cell surface binding site for *Clostridium difficile* enterotoxin: evidence for a glycoconjugate containing the sequence Galα1-3Galβ1-4GlcNAc. Infect Immun 53:573–581

Libby JM, Jortner BS, Wilkins TD (1982) Effects of the two toxins of *Clostridium difficile* in antibiotic-associated cecitis in hamsters. Infect Immun 36:822–829

Lima AAM, Lyerly DM, Wilkins TD, Innes DJ, Guerrant RL (1988) Effects of *Clostridium difficile* toxins A and B in rabbit small and large intestine in vivo and on cultured cells in vitro. Infect Immun 56: 582–588

Lyerly DM, Lockwood DE, Richardson SH, Wilkins TD (1982) Biological activities of toxins A and B of *Clostridium difficile*. Infect Immun 35:1147–1150

Lyerly DM, Sullivan NM, Wilkins TD (1983) Enzyme-linked immunosorbent assay for *Clostridium difficile* toxin A. J Clin Microbiol 17:72–78

Lyerly DM, Saum KE, MacDonald DK, Wilkins TD (1985) Effects of *Clostridium difficile* toxins given intragastrically to animals. Infect Immun 47:349–352

Lyerly DM, Roberts MD, Phelps CJ, Wilkins TD (1986) Purification and properties of toxins A and B of *Clostridium difficile*. FEMS Microbiol Lett 33:31–35

Lyerly DM, Barroso LA, Wilkins TD, Depitre C, Corthier G (1992) Characterization of a toxin A-negative, toxin B-positive strain of *Clostridium difficile*. Infect Immun 60:4633–4639

Lyerly DM, Wilkins TD (1995) *Clostridium difficile*. In: Blaser MJ, Smith PD, Ravdin JI, Greenberg HB, Guerrant RL (eds) Infections of the Gastrointestinal Tract. Raven Press, New York, pp 867–891

Makay DJG, Hall A (1998) Rho GTPases. J Biol Chem 273:20685–20688

Manyth CR, Maggio JE, Mantyh PW, Vigna SR, Pappas TN (1996a) Increased substance P receptor expression by blood vessels and lymphoid tissue aggregates in *Clostridium difficile*-induced pseudo-membranous colitis. Dig Dis Sci 41:614–620

Mantyh CR, Pappas TN, Lapp JA, Washington MK, Neville LM, Ghilardi JR, Rogers SD, Manyth PW, Vigna SR (1996b) Substance P activation of enteric neurons in response to intraluminal *Clostridium difficile* toxin A in the rat ileum. Gastroenterology 111:1272–1280

Martinez RD, Wilkins TD (1988) Purification and characterization of *Clostridium sordellii* hemor-rhagic toxin and cross-reactivity with *Clostridium difficile* toxin A (enterotoxin). Infect Immun 56: 1215–1221

Martinez RD, Wilkins TD (1992) Comparison of *Clostridium sordellii* toxins HT and LT with toxins A and B of *C. difficile*. J Med Microbiol 36:30–36

Marvaud JC, Eisel U, Binz T, Nieman H, Popoff MR (1998) TetR is a positive regulator of the tetanus toxin gene in *Clostridium tetani* and is homologous to botR. Infect Immun 66:5698–5702

Marvaud JC, Gibert M, Inoue K, Fujinaga Y, Oguma K, Popoff MR (1998) botR/A is a positive regulator of the botulinum neurotoxin and associated non-toxin protein genes in *Clostridium botulinum* A. Mol Microbiol 29:1009–1018

Missiakas D, Raina S (1998) The extracytoplasmic function factors: role and regulation. Mol Microbiol 28:1059–1066

Mitchell J, Laughon BE, Lin S (1987) Biochemical studies on the effect of *Clostridium difficile* toxin B on actin in vivo and in vitro. Infect Immun 55:1610–1615

Moncrief JS, Barroso LA, Wilkins TD (1997a) Positive regulation of *Clostridium difficile* toxins. Infect Immun 65:1105–1108

Moncrief JS, Lyerly DM, Wilkins TD (1997b) Molecular biology of the *Clostridium difficile* toxins. In: Rood JI, Songer G, McClane B, Titball R (eds) Molecular Genetics and Pathogenesis of the Clostridia. Academic Press, London, pp 369–392

Moncrief JS, Duncan AJ, Wright RL, Barroso LA, Wilkins TD (1998) Molecular characterization of the fragilysin pathogenicity islet of enterotoxigenic *Bacteroides fragilis*. Infect Immun 66:1735–1739

Moncrief JS, Neville LM, Lyerly DM (1999) Molecular characterization of toxin A-negative/toxin B-positive strains of *Clostridium difficile*. Am Soc Microbiol, Abstr Ann Meet, 99th

Muldrow LL, Ibeanu GC, Lee NI, Bose NK, Johnson J (1987) Molecular cloning of *Clostridium difficile* toxin A gene fragment in lambda/gt11. FEBS Lett 213:249–253

Mullany P, Wilks M, Lamb I, Clayton C, Wren B, Tobaqchali S (1990) Genetic analysis of a tetracycline resistance element from *Clostridium difficile* and its conjugal transfer to and from *Bacillus subtilis*. J Gen Microbiol 136:1343–1349

Mullany P, Wilks M, Tobaqchali S (1991) Transfer of Tn*916* and Tn*916*ΔE into *Clostridium difficile*: demonstration of a hot spot for these elements in the *C. difficile* genome. FEMS Microbiol Lett 79:191–194

Mullany P, Wilks M, Puckey L, Tabaqchali S (1994) Gene cloning in *Clostridium difficile* using Tn*916* as a shuttle conjugative transposon. Plasmid 31:320–323

Ottlinger ME, Lin S (1988) *Clostridium difficile* toxin B induces reorganization of actin, vinculin and talin in cultured cells. Exp Cell Res 174:215–229

Phelps CJ, Lyerly DM, Johnson JL, Wilkins TD (1991) Construction and expression of the complete *Clostridium difficile* toxin A gene in *Escherichia coli*. Infect Immun 59:150–153

Popoff MR (1986) Purification and characterization of *Clostridium sordellii* lethal toxin and cross-reactivity with *Clostridium difficile* cytotoxin. Infect Immun 55:35–43

Popoff MR, Chaves-Olarte E, Lemichez E, von Eichel-Streiber C, Thelstam M, Chardin P, Cusssac D, Antony B, Chavier P, Flatau G, Giry M, de Gunzburg J, Boquet P (1996) Ras, Rap, and Rac small GTP-binding proteins are targets for *Clostridium sordellii* lethal toxin glucosylation. J Biol Chem 271:10217–10224

Pothoulakis C, Castigiuolo I, LaMont JT, Jaffer A, O'Keane JC, Snider RM, Leeman SE (1994) CP-96,345, a substance P antagonist inhibits rat intestinal responses to *Clostridium difficile* toxin A but not cholera toxin. Proc Natl Acad Sci USA 91:947–951

Pothoulakis C, Gilbert RJ, Cladaras C, Castagliuolo I, Semenza G, Hitti Y, Moncrief JS, Linevsky J, Kelly CP, Nikulasson S, Desai HP, Wilkins TD, LaMont JT (1996) Rabbit sucrase-isomaltase contains a functional intestinal receptor for *Clostridium difficile* toxin A. J Clin Invest 98:641–649

Pothoulakis C, Castagliuolo I, Leeman SE, Wang CC, Li H, Hofmann BJ, Mezey E (1998) Substance P receptor expression in intestinal epithelium in *Clostridium difficile* toxin A enteritis in rats. Am J Physiol 275:G68–G75

Price SB, Phelps CJ, Wilkins TD, Johnson JL (1987) Cloning of the carbohydrate binding portion of the toxin A gene of *Clostridium difficile*. Curr Microbiol 16:55–60

Riegler M, Sedivy R, Pothoulakis C, Hamilton G, Johannes Z, Bischof G, Costentini E, Wolfgang F, Schiessel, LaMont JT, Wenzl E (1995) *Clostridium difficile* toxin B is more potent than toxin A in damaging human colonic epithelium in vitro. J Clin Invest 95:2004–2011

Rupkin M, Braun V, Soehn F, Janc M, Hofstetter M, Laufenberg-Feldmann R, von Eichel-Streiber C (1997) Characterization of polymorphisms in the toxin A and B genes of *Clostridium difficile*. FEMS Microbiol Lett 148:197–202

Rupkin M, Avesani V, Janc M, Eichel-Streiber Cv, Delmee M (1998) A novel toxinotyping scheme and correlation of toxinotypes with serogroups of *Clostridium difficile* isolates. J Clin Microbiol 36:2240–2247

Sehr P, Joseph G, Genth H, Just I, Pick E, Aktories K (1998) Glucosylation and ADP ribosylation of rho proteins: effects on nucleotide binding, GTPase activity, and effector coupling. Biochemistry 37:5296–5304

Selzer J, Hofmann F, Rex G, Wilm M, Mann M, Just I, Aktories K (1996) *Clostridium novyi* alpha-toxin-catalyzed incorporation of GlcNAc into Rho subfamily proteins. J Biol Chem 271:25173–25177

Smith JA, Cooke DL, Hyde S, Borrielo SP, Long (1997) *Clostridium difficle* binding to human epithelial cells. J Med Microbiol 46:953–958

Soehn F, Wagenknecht-Wiesner A, Leukel P, Kohl M, Weidmann M, von Eichel-Streiber C, Braun V (1998) Genetic rearrangements in the pathogenicity locus of *Clostridium difficile* strain 8864 – implications for the transcription, expression and enzymatic activity of toxin A and B. Mol Gen Genet 258:222–232

Song KP, Faust C (1998) Molecular analysis of the promoter region of the *Clostridium difficile* toxin B gene that is functional in *Escherichia coli*. J Med Microbiol 47:309–316

Sterne M, Wentzel LM (1950) A new method for the large scale production of high titre botulinum Formal-toxoid types C and D. J Immunol 65:175–183

Thelestam M, Bronnegard M (1980) Interaction of cytopathogenic toxin from *Clostridium difficile* with cells in tissue culture. Scand J Infect Dis 22:16–29

Torres JF (1991) Purification and characterization of toxin B from a strain of *Clostridium difficile* that does not produce toxin A. J Med Microbiol 35:40–44

Tucker KD, Carrig PE, Wilkins TD (1990) Toxin A of *Clostridium difficle* is a potent cytotoxin. J Clin Microbiol 28:869–871

Tucker K, Wilkins TD (1991) Toxin A of *Clostridium difficle* binds to the human carbohydrate antigens I, X, and Y. Infect Immun 59:73–78

Wagenknecht-Wiesner A, Weidmann M, Braun V, Leukel P, Moos M, von Eichel-Streiber Cv (1997) Delineation of the catalytic domain of the *Clostridium difficile* toxin B-10463 to an enzymatically active N-terminal 467 amino acid fragment. FEMS Microbiol Lett 152:109–116

Wren BW, Clayton CL, Mullany PP, Tabaqchali S (1989) Molecular cloning and expression of *Clostridium difficile* toxin A in *Escherichia coli* K 12. FEBS Lett 225:82–86

Yamakawa K, Karasava T, Ikoma S, Nakamura S (1996) Enhancement of *Clostridium difficile* toxin production in biotin-limited conditions. J Med Microbiol 44:111–114

Molecular Mode of Action
of the Large Clostridial Cytotoxins

I. JUST, F. HOFMANN, and K. AKTORIES

1 Introduction

The large clostridial cytotoxins are a family of functionally and structurally related toxins produced by clostridia comprising *Clostridium difficile* toxin A and toxin B, *Clostridium sordellii* lethal and haemorrhagic toxin and *Clostridium novyi* α-toxin (Table 1). These toxins are exotoxins which induce morphological changes of the cultured target cells based on the redistribution of the microfilament system. The cytotoxic activity on cultured cell lines led to their designation as cytotoxins. Despite their comparable in vitro effects, the cytotoxins – as major pathogenicity factors – are involved in different diseases and clinical outcomes. *Clostridium difficile* toxins A and B are of major clinical importance because both toxins are the causative agents in about 20% of antibiotic-associated diarrhoea and in almost all

Institut für Pharmakologie und Toxikologie der Universität Freiburg, Hermann-Herderstrasse 5, 79104 Freiburg, Germany

Table 1. Protein substrates and cosubstrates of the large clostridial cytotoxins

Toxin	Cosubstrate	Transferred moiety	Substrates and site of modification ( )
Clostridium difficile toxin A (308kDa)	UDP-glucose	Glucose	RhoA(T37), Rac(T35), Cdc42(T35), Rap(T35)
Clostridium difficile toxin B (270kDa)	UDP-glucose	Glucose	RhoA(T37), Rac(T35), Cdc42(T35)
Clostridium sordellii lethal toxin (271kDa)	UDP-glucose	Glucose	Rac(T37), Cdc42(T35), Ras(T35), Rap(T35), RalA(T46)[a]
Clostridium sordellii haemorrhagic toxin (∼300kDa)	UDP-glucose	Glucose	RhoA(T37), Rac(T35), Cdc42(T35)
Clostridium novyi α-toxin (250kDa)	UDP-*N*-acetyl-glucosamine	*N*-acetyl-glucosamine	RhoA(T37), Rac(T35), Cdc42(T35), RhoG(T35)

T, threonine.
[a] The protein substrate specificity depends on the strain of *Clostridium sordellii*.

cases of pseudomembranous colitis (KELLY et al. 1994; KELLY and LAMONT 1998; BARTLETT 1994).

Because toxin A, but not toxin B, induces fluid accumulation in animal models and ileal explants it was designated as enterotoxin (TRIADAFILOPOULOS et al. 1987; LYERLY et al. 1985). Toxin B, in contrast, is about 100 to 1000-fold more cytotoxic to cultured cell lines and has therefore been named a cytotoxin (LYERLY et al. 1982). Both toxins induce the same morphological and cytoskeletal features, their cytotoxic activity differs only in potency. The cytotoxic outcome is characterised by shrinking and rounding of cells, initially accompanied by formation of neurite-like retraction fibres. In the progress of intoxication, the retraction fibres disappear and the cells completely round up, while in the terminal phase, the cells partially detach. The morphological changes are accompanied by a redistribution of the actin cytoskeleton. The cell-spanning stress fibres disappear and the remainder of the actin filaments accumulate in the perinuclear space (SIFFERT et al. 1993; FIORENTINI et al. 1989, 1990, 1993; MALORNI et al. 1991; CIESIELSKI-TRESKA et al. 1989). Other members of the family of large clostridial cytotoxins induce the same effects with some variations in detailed aspects (BETTE et al. 1991; OKSCHE et al. 1992; POPOFF 1987; CIESIELSKI-TRESKA et al. 1991).

2 Design of the Cytotoxins

The cytotoxins are single-chained peptides with molecular masses between 250 to 308kDa encompassing three functional domains (Fig. 1; Fig. 2 gives the alignment of the cytotoxins). The enzymatic domain which catalyses the monoglucosylation reaction is located N-terminally (HOFMANN et al. 1997). The intermediary part [at about amino acid (aa) 1000] encloses a predicted transmembrane domain, which is likely to mediate the translocation of the toxin into the cytosol, but this mechanism

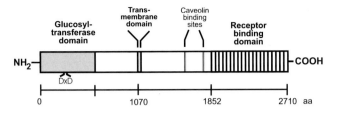

Fig. 1. Structure of *Clostridium difficile* toxin A. The structure of toxin A is depicted as an example for the homologous large clostridial cytotoxins. The toxins of three functional domains are designed. The receptor-binding domain is located at the C-terminal part composed of repetitive peptide elements. The transmembrane domain located in the intermediate part is supposed to mediate the translocation of the toxin or a toxin fragment from acidic endosomal compartments into the cytosol. The N-terminal part covers the glucosyltransferase activity. The DXD motif is part of the catalytic domain responsible for the binding of Mn^{2+}

has not been proven yet. The receptor-binding domain resides in the C-terminally located third of the molecule which is constructed of repetitive peptide elements. Three findings have led to the notion that this part really covers the receptor-binding domain: (a) the modular organisation of this region (WREN 1991; VON EICHEL-STREIBER et al. 1992b); (b) the homology to the carbohydrate-binding domain of streptococcal glycosyltransferases (VON EICHEL-STREIBER et al. 1992a,b); and (c) the inhibition of the cytotoxic activity of toxin A by a monoclonal antibody recognising an epitope of this domain (LYERLY et al. 1986; FREY and WILKINS 1992) and inhibition of *Clostridium difficile*-associated morbidity in an animal model through antibodies directed to the C-terminal part of toxin A and toxin B (KINK and WILLIAMS 1998). However, the finding that deletion of the C-terminal repetitive domain of toxin B decreases the cytotoxic potency only by a factor of ten is not reconcilable with this concept (BARROSO et al. 1994).

Toxin A and B possess two consensus sequences for caveolin binding (toxin A, amino acids 1595–1602 and 1791–1800; toxin B, amino acids 657–666 and 1257–1266). Besides involvement in the formation of caveolae (LISANTI et al. 1994), caveolin is also a scaffolding protein forming signalling complexes for heterotrimeric G-proteins, H-Ras and Src kinases (LISANTI et al. 1994; OKAMOTO et al. 1998). Cholera toxin which ADP-ribosylates Gα is thought to be directed to membranous microdomains through its caveolin-binding site to get into close contact with its protein substrate (LISANTI et al. 1994; COUET et al. 1997). It is conceivable that the intracellular localisation of toxin A and B is determined through their putative caveolin-binding sites.

3 Cell Admission

Toxin A and B are intracellularly acting cytotoxins which get access to their target cells via receptor-mediated endocytosis. Cultured cell lines show saturable binding

```
             10        20        30        40        50        60        70
ToxinB  MSLVNRKQLEKMANVRFRTQEDEYVAILDALEEYHN-MSENTVVEKYLKLKDINSLTDIYIDTYKKSGRN
LT      MNLVNKAQLQKMVYVKFRIQEDEYVAILNALEEYHN-MSESSVVEKYLKLKDINNLTDNYLNTYKKSGRN
ToxinA  MSLISKEELIKLAYS-IRPRENEYKTILTNLDEYNK-LTTNNNENKYLQLKKLNESIDVFMNKYKTSSRN
ALPHA   -MLITREQLMKIASIPLKRKEPEYNLILDALENFNRDIEGTSVKEIYSKLSKLNELVDNYQTKYPSSGRN

             80        90       100       110       120       130       140
ToxinB  KALKKFKEYLVTEVLEL-KNNNLTPV-EKNLHFWIGGQINDTAINYINQWKDVNSDYNVNVFYDSNAFL
LT      KALKKFKEYLTMEVLEL-KNNSLTPV-EKNLHFIWIGGQINDTAINYINQWKDVNSDYTVKVFYDSNAFL
ToxinA  RALSNLKKDILKEVILI-KNSNTSPV-EKNLHFVWIGGEVSDIALEYIKQWADINAEYNIKLWYDSEAFL
ALPHA   LALENFRDSLYSELRELIKNSRTSTIASKNLSFIWIGGPISDQSLEYYNMWKMFNKDYNIRLFYDKNSLL

            150       160       170       180       190       200       210
ToxinB  INTLKKTVVESAINDTLESFRENLNDPRFDYNKFFRKRMEIIYDKQKNFINYYKAQREENPELIIDDIVK
LT      INTLKKTIVESATNNTLESFRENLNDPEFDYNKFYRKRMEIIYDKQKHFIDYYKSQIEENPEFIIDNIIK
ToxinA  VNTLKKAIVESSTTEALQLLEEEIQNPQFDNMKFYFVDRQKRFINYYKSQINKPTVPTIDDIIK
ALPHA   VNTLKTAIIQESSKVIIEQNQSNILDGTYGHNKFYSDRMKLIYRYKRELKMLYENMKQNNS---VDDIII

            220       230       240       250       260       270       280
ToxinB  TYLSNEYSKEIDELNTYIEESLNKITQNSGNDVRNFEEFKNGESFNLYEQELVERWNLAAASDILRISAL
LT      TYLSNEYSKDLEALNKYIEESLNKITANNGNDIRNLEKFADEDLVRLYNQELVERWNLAAASDILRISML
ToxinA  SHLVSEYNRDETVLESYRTNSLRKINSNHGIDIRANSLFTEQELLNIYSQELLNRGNLAAASDIVRLLAL
ALPHA   NFLSNYFKYDIGKLNNQKENNNKMAIGATDINTENILTN-KLKSYYQELIQTNNLAAASDILRIAIL
```

Fig. 2. Alignment of large clostridial cytotoxin-deduced amino acid sequences. Toxin B from *Clostridium difficile*; lethal toxin (*LT*) from *Clostridium sordellii*; toxin A from *Clostridium difficile*; α-toxin (*ALPHA*) from *Clostridium novyi*. Sequences were aligned using Clustal W (THOMPSON et al. 1994) (multiple alignment parameters, weight matrix = blosum; gap opening penalty = 10.0; gap extension penalty = 0.05). Identity, |; high similarity, *

```
                290       300       310       320       330       340       350
                 -         -         -         -         -         -         -
ToxinB   KEIGGMYLDVDMLPGIQPDLFESIEKPSSVTVDFWEMTKLEAIMKYKEYIPEYTSEHFDMLDEEVQSSFE
LT       KEDGGVYLDVDILPGIQPDLFKSINKPDSITNTSWEMIKLEAIMKYKEYIPGYTSKNFDMLDEEVQRSFE
ToxinA   KNFGGVYLDVDMLPGIHSDLFKTISRPSSIGLDRWEMIKLEAIMKYKYKYINNYTSENFDKLDQQLKDNFK
ALPHA    KKYGGVYCDLDFLPGVNLSLFNDISKPNGMDSNYWEAAIFEAIANEKKLMNNYPYKYMEQVPSEIKERIL
         |* ||*| |*|*|||**   ||*     ||    *  | *    | ***   *  |  *  *  ***

                360       370       380       390       400       410       420
                 -         -         -         -         -         -         -
ToxinB   SVLASKSDKSEIFSSLGDMEASPLEV----KIAFNSKGIINQGLISVKDSYCSNLIVKQIENRYKILNN
LT       SALSSKSDKSEIFLPLDDIKVSPLEV----KIAFANNSVINQALISLKDSYCSDLVINQIKNRYKILND
ToxinA   LIIESKSEKSEIFSKLENLNVSDLEI-----KIAFALGSVINQALISKQGSYLTNLVIEQVKNRYQFLNQ
ALPHA    SFVRNH-DINDLILPLGDIKISQLEILLSRLKAATGKKTFSNAFIISNNDSLTNNLISQLENRYEILNS
           *  *    *** |   ||*         *        *||  *||    |*||**||   *

                430       440       450       460       470       480       490
                 -         -         -         -         -         -         -
ToxinB   SLNPAI---SEDNDFNTTTNTFIDSIMAEANADNGRFMMELGKYLRVGFFPDVKTTINLSGPEAYAAAY
LT       NLNPSI----NEGTDFNTTMKIFSDKLASISNEDNMMFMIKITNYLKVGFAPDVRSTINLSGPGVYTGAY
ToxinA   HLNPAI----ESDNNFTDTTKIFHDSLFNSATAENSMFLTKIAPYLQVGFMPEARSTISLSGPGAYASAY
ALPHA    IIQEKFKICETYDSYINSVSELVLETTPKNLSMDGSSFYQQIGYLSSGFKPEVNSTVFFSGPNIYSSAT
          **    *            *                 | **     ||| |*  |*

                500       510       520       530       540       550       560
                 -         -         -         -         -         -         -
ToxinB   QDLLMFKEGSMNIHLIEADLRNFEISKTNISQSTEQEMASLWSFDDARAKAQFEEYKRNYFEGSLGEDDN
LT       QDLLMFKDNSTNIHLLEPELRNFEFPKTKISQLTEQEITSLWSFNQARAKSQFEEYKKGYFEGALGEDDN
ToxinA   YDFINLQENTIEKTLKASDLIEFKFPENNLSQLTEQEINSLWSFDQASAKYQFEKYVRDYTGGSLSEDNG
ALPHA    CDTYHFIKNTFDM-LSSQNQEIFEASNNLYFSKTHDEFKSSWLLRSNIAEKEFQKLIKTYIGRTLNYEDG
          |            |*  | *|*      *  |* *|* *|** *   * |*
```

Fig. 2. (*Contd.*)

```
               570       580       590       600       610       620       630
ToxinB  LDFSQNIVDKEYLLEKISSLARSSE---RGYIHYIVQLQGDKISYEAAACNLFAKTPYDSVLFQKNIEDS
LT      LDFAQNTVLDKDYVSKKILSSMKTRN--KEYIHYIVQLQGDKISYEASCNLFSKDPYSSILYQKNIEGS
ToxinA  VDFNKNTALDKNYLLNNKIPSNNVEEAGSKNYVHYIIQLQGDDISYEATCNLFSKNPKNSIIQRNMNES
ALPHA   LNFNKWKRVTTSELLKVIEEVNSTKIY-ENYDLNMILQIQGDDISYESAVNVFGKNPNKSILIQ-GVDDF
        **|  *          *   *           **  |*|*|||  |||*|**  |*|   |**

               640       650       660       670       680       690       700
ToxinB  EIAYYNPGDGEIQEIDKYKIPSIISDRPKIKLTFIGHGKDEFNTDIFAGFDVDSLSTEI-----EAAID
LT      ETAYYYVADAEIKEIDKYRIPYQISNKRNIKLTFIGHGKSEFNTDTFANLDVDSLSSEI-----ETILN
ToxinA  AKSYFLSDDGESILELNKYRIPERLKNKEKVKVTFIGHGKDEFNTSEFARLSVDSLSNEI----SSFLD
ALPHA   ANVFYFENGIVQSDNIN--NILSRFNDIKKIKLTLIGHGENVFNPKLFGGKTVNDLYTNIIKPKLQHLLE
        **        ***|      *     **|*|||||*||   |  | * |   |*|    **

               710       720       730       740       750       760       770
ToxinB  LAKEDISPKSIEINLLGCNMFSYSINVEETYPGKLLLKVKDKISELMPSISQDSIIVSANQYEVRINSEG
LT      LAKADISPKYIEINLLGCNMFSYSISAEETYPGKLLLKIKDRVSELMPSISQDSITVSANQYEVRINEEG
ToxinA  TIKLDISPKNVEVNLLGCNMFSYDFNVEETYPGKLLLSIMDKITSTLPDVNKNSITIGANQYEVRINSEG
ALPHA   REGVILKNKYLKINILGCYMFTPKVDINSTFVGKLFNKISRDLQP--KGFSKNQLEISANKYAIRINREG
        * |  ***|*|||  ||*  | ||  ||*| | **  *   |||*

               780       790       800       810       820       830       840
ToxinB  RRELLDHSGEWINKEESIIKDISSKEYISFNPKENKITVKSKNLPELSTLLQEIRNNSNSSDIELEEKVM
LT      KREILDHSGKWINKEESIIKDISSKEYISFNPKENKIIVKSKYLHELSTLLQEIRNNANSSDIDLEKKVM
ToxinA  RKELLAHSGKWINKEEAIMSDLSSKEYIFFDSIDNKLKAKSKNIPGLASISEDIKTLLLDASVSPDTKFI
ALPHA   KREVLDYFGKWVSNTDLIAEQISNKYVVYWNEVENTLSARVEQLNKVAEFAKDINSIIQTTN-NQELKQS
        **|*|*  |*|*  *     ** |  |* **||||  ** *  ***
```

Fig. 2. (*Contd.*)

```
            850       860       870       880       890       900       910

ToxinB  LTECEINVISNIDTQIVEERIE-EAKNLTSDSINYIKDEFKLIESISDALCDLKQQNELEDSHFISFEDI
LT      LTECEINVASNIDRQIVEGRIE-EAKNLTSDSINYIKNEFKLIESISDSLYDLKHQNGLDDSHFISFEDI
ToxinA  LNNLKLNIESSIGDYIYYEKLE-PVKNIIHNSIDDLIDEFNLLENVSDELYELKKLNNLDEKYLISFEDI
ALPHA   LVNTYADLITTLYSELLKEDIPFFELDNIQIKE-RIILNEISRLHDFSNIILDFYQKNNISNNMILLFDSI
        | *    * *  *** * *  |     *  * *** * * *** |  | * *  |*  | |*  |


            920       930       940       950       960       970       980

ToxinB  SETDEGFSIRFINKETGESIFVETEKTIFSEYANHITEEISKIKGTIFDTVNGKLVKKVNLDTTHEVNTL
LT      SKTENGFRIRFINKETGNSIFIETEKEIFSEYATHISKEISNIKDTIFDNVNGKLVKKVNLDAAHEVNTL
ToxinA  SKNNSTYSVRFINKSNGESVYVETEKEIFSKYSEHITKEISTIKNSIITDVNGNLLDNIQLDHTSQVNTL
ALPHA   IKEKDYYNVKLANKITGETSVIKTYSDSLWNFTNKYKKIVDDIKGIIVKDINGEFIKKADFEIEQNPSLL
        *  ***   |    |  |  ** **** *           *    *|| *|**** ** * *** *


            990      1000      1010      1020      1030      1040      1050

ToxinB  NAAFFIQSLIEYNSSKESLSNLSVAMKVQVYAQLFSTGLNTITDAAKVVELVSTALDETIDLLPTLSEGL
LT      NSAFFIQSLIEYNTTKESLSNLSVAMKVQVYAQLFSTGLNTITDASKVVELVSTALDETIDLLPTLSEGL
ToxinA  NAAFFIQSLIDYSSNKDVLNDLSTSVKVQLYAQLFSTGLNTIYDSIQLVNLISNAVNDTINVLPTITEGI
ALPHA   NSAMLMQLLIDYKPYTEILTNMNTSLKVQAYAQIFQLSIGAIQEATEIVTIISDALNANFNILSKLKVGS
        |*|***  | |*|      |  ** ***  *   **|  | | *** | ***|  |*      * ***


           1060      1070      1080      1090      1100      1110      1120

ToxinB  PIIATIIDGVSLGAAIKELSETSDPLLRQEIEAKIGIMAVNLTTATTAIITSSLGIASGFSILLV---PL
LT      PIIATIIDGVSLGAAIKELSETNDPLLRQEIEAKIGIMAVNLTAASTAIVTSALGIASGFSILLV---PL
ToxinA  PIVSTILDGINLGAAIKELLDEHDPLLKKELEAKVGVLAINMSLSIAATVASIVGIGAEVTIFLL---PI
ALPHA   SVASVIIDGINLIAALTELKNVKTNFERKLIEAKVGMYSIGFILESSSLISGLLGATAVSEILGVISVPV
        * * *||* || || |    * * ** ***|*|** *   * ***  *|      | |* |*  |*
```

Fig. 2. (Contd.)

```
             1130      1140      1150      1160      1170      1180      1190
ToxinB  AGISAGIPSLVNNELVLRDKATKVVDYFKHVSLVETEGVFTLLDDKIMPQDDLVISEIDFNNNSIVLGK
LT      AGISAGIPSLVNNELILQDKATKVIDYFKHISLAETEGAFTLLDDKIIMPQDDLVLSEIDFNNNSITLGK
ToxinA  AGISAGIPSLVNNELILHDKATSVVNYFNHLSESKKYGPLKTEDDKILVPIDDLVISEIDFNNNSIKLGT
ALPHA   AGILVGLPSLVNNILVLGEKYNQILDYFSKFYPIVGKNPFSIQDN-IIIPYDDIAITELNFKYNKFKYGY

             1200      1210      1220      1230      1240      1250      1260
ToxinB  CEIWRMEGGSGHTVTDDIDHFFSAPSITYREPHLSIYDVLEVQKEELDLSKDLMVLPNAPNRVFAWETGW
LT      CEIWRAEGGSGHTLTDDIDHFFSSPSITYRKPWLSIYDVLNIKKEKIDFSKDLMVLPNAPNRVFGYEMGW
ToxinA  CNILAMEGGSGHTVTGNIDHFFSSPSISSHIPSLSIYSAIGIETENLDFSKKIMMLPNAPSRVFWWETGA
ALPHA   AKISGLKVGLVTHIGENIDHYFSAPSLDHYIE-LSIYPALKLNDTNLP-KGNVVLLPSGLNKVYKPEISA

             1270      1280      1290      1300      1310      1320      1330
ToxinB  TPGLRSLENDGTKLLDRIRDNY---EG--EFYWRYFAFIADALITTLKPRYEDTNIRINLDSNTRSFIVP
LT      TPGFRSLDNDGTKLLDRIRDHY---EG--QFYWRYFAFIADALITKLKPRYEDTNVRINLDGNTRSFIVP
ToxinA  VPGLRSLENDGTRLLDSIRDLY---PG--KFYWRFYAFF-DYAITTLKPVYEDTNIKIKLDKDTRNFIMP
ALPHA   IAGANSQEGNGVEVLNLIRNYYVDSNGNTKFPWKYEAPF-EYSFSYMRVEYFDTKVNVILDNENKTLIIP

             1340      1350      1360      1370      1380      1390      1400
ToxinB  IITTEYIREKLSYSFYGSGGTYALSLSQYNMGINIELSESDVWIIDVDNVVRDVTIESDKIKKGDLIEGI
LT      VITTEQIRKNLSYSFYGSGGSYSLSLSPYNMNIDLNLVENDTWVIDVDNVKNITIESDEIQKGELIENI
ToxinA  TITTNEIRKNLSYSFDGAGGTYSLLLSSYPISTNINLSKDDLWIFNIDNEVREISIENGTIKKGKLIKDV
ALPHA   VLTIDEMRNKISYEILGDGQYNVILPVNQTNINIVSNKNDIWNFDVSYIVKESKIEDNKFVLDGFINNI
```

Fig. 2. (*Contd.*)

```
              1410      1420      1430      1440      1450      1460      1470
ToxinB  LSTLSIEENKIILNSHEINFSGEVNGSNGFVSLTFSILEGINAIIEVDL-LSKSYKL-LISGELKILMLN
LT      LSKLNIEDNKIILNNHTINFYGDINESNRFISLTFSILEDINIIIEIDL-VSKSYKI-LLSGNCMKLIEN
ToxinA  LSKIDINKNKLIIGNQTIDFSGDIDNKDRYIFLTCELDDKISLIIEINL-VAKSYSL-LLSGDKNYLISN
ALPHA   FSTLKVSNDGFKIGKQFIS----IKNTPRAINLSFKINNNI-VIVSIYLNHEKSNSITIISSDLNDIKNN

              1480      1490      1500      1510      1520      1530      1540
ToxinB  SNHIQQKIDYIGFNSELQKNIPYSFVD-SEGKENGFINGSTKEGLFVSELPDVVLISKVYMDDSKPSFGY
LT      SSDIQQKIDHIGFNGEHQKYIPYSYID-NETKYNGFIDYSKKEGLFTAEFSNESIIRNIYMPDSNNLFIY
ToxinA  LSNTIEKINTLGLDS--KNIAYNYTDESNNKYFGAISKTSQKSIIHYKKDSKNILE--FYNDSTLEFNS
ALPHA   FDNLLDNINYIGLGS-ISDNTINCIVRNDEVYMEGKIFLNEKKLVFIQNELELHLYDS-VNKDSQYLINN

              1550      1560      1570      1580      1590      1600      1610
ToxinB  YSNNLKDVKVITKDNVNILTGYYLKD---DIKISLSLTLQDEKTIKLNSVHLDESGVAEILKFMNR-KGN
LT      SSKDLKDIRIINKGDVKLLIGNYFKD---DMKVSLSFTIEDTNTIKLNGVYLDENGVAQILKFMNNAKSA
ToxinA  KDFIAEDINVFMKDDINTITGKYYVDNNTDKSIDFSISLVSKNQVKVNGLYLNESVYSSYLDFVKNSDGH
ALPHA   PIN-----NVVKYKDGYIVEGTFLIN---STENKYSLYIEN-NKIMLKGLYLESSVFKTIQDKIYS---K

              1620      1630      1640      1650      1660      1670      1680
ToxinB  TNTSDSLMSFLESMNIKSIFVNFLQSNIKFILDANFIISG-TTSIGQFEFICDENDNIQPYFIKFNTLET
LT      LNTSNSLMNFLESINIKNIFYNNLDPNIEFILDTNFIISG-SNSIGQFELICDKDKNIQPYFINFKIKET
ToxinA  HNTSNFMNLFLDNISFWKLF-GF--ENINFVIDKYFTLVG-KTNLGYVEFICDNNKNIDIYFGEWKTSSS
ALPHA   EKVNDYILS-L--I--KKFF------TVNIQLCPFMIVSGVDENNRYLEYMLSTNNKWIINGGYWENDFN
```

Fig. 2. (*Contd.*)

```
             1690      1700      1710      1720      1730      1740      1750

ToxinB  NYTLYVGNRQNMIVEPNYDLDDSGDISSTVINFSQKYLYGIDSCVNKVVISPNIYTDEINITPVYETNNT  —
LT      SYTLYVGNRQNLIVEPSYHLDDSGNISSTVINFSQKYLYGIDRYVNKVIIAPNLYTDEINITPVYKPNYI  —
ToxinA  KSTIFSGNGRNVVEPIYNPDTGEDISTS-LDFSYEPLYGIDRYINKVLIAPDLYTSLININTNYYSNEY   —
ALPHA   NYKIVDFEKCNVIVSGSNKLNSEGDLADT-IDVLDKDLENL--YIDSVIIPKVYTKKIIIHPI--PNN-   —
          *   |**|  *  *** * ** **** *  *** **  *  |*||  |  *|  —

             1760      1770      1780      1790      1800      1810      1820

ToxinB  YPEVIVLDANYINEKINVNIND-LSIRYVWSNDGNDFILMSTSEENKVSQV-KIRFVNVFKDKTLANKLS  —
LT      CPEVIILDANYINEKINVNIND-LSIRYVWDNDGSDLILIANSEEDNQPQV-KIRFVNVFKSDTAADKLS  —
ToxinA  YPEIIVLNPNTFHKKVNINLDS-SSFEYKWSTEGSDFILVRYLEESNKKILQKRIKGILSNTQSFNKMS   —
ALPHA   -PQINIINTQSIHDKCHLIIDSVLTNNYHWESDGDDLI-ITNGLDINIRIL----------QGLS       |
         |** **** *  **  *  |  ** *   *   *|  *          *     * *|

             1830      1840      1850      1860      1870      1880      1890

ToxinB  FNFSDKQDVPVSEIILSFTPSYYEDGLIGYDLGLVSLYNEKFYINNFGMMVSGLIYINDSLYYFKPPVNN  —
LT      FNFSDKQDVDSVSKIISTFSLAAYSDGFFDYEFGLVSLDNDYFYINSFGNMVSGLIYINDSLYYFKPPKNN  —
ToxinA  IDFKDIKKLSLGYIMSNFKSFNSENELDRDHLGFKIIDNKTYYDEDSKLVKGLININNSLFYFDPIEFN   —
ALPHA   FGFKYK-N-----IYLKFSNY------DELSL----ND-FLLQNYN--VKGLYYINGELHYKNIPGDT    —
         *   |                               *    *

             1900      1910      1920      1930      1940      1950      1960

ToxinB  LITGFVTVGDDKYYFNPINGGAASIGETIIDDKNYYFN-QSGVLQTGVFSTEDGFKYFAPANTLDENLEG  —
LT      LITGFTTIDGNKYYFDPTKSGAASIGEITIDGKDYYFN-KQGILQVGVINTSDGLKYFAPAGTLDENLEG  —
ToxinA  LVTGWQTINGKKYYFD-INTGAALTSYKIINGKHFYFN-NDGVMQLGVFKGPDGFEYFAPANTQNNNIEG  —
ALPHA   FEYGWINIDSRWYFFD-SINLIAKKGYQEIEGERYYFNPNTGVQESGVFLTPNGLEYFTNKHASSKRW-G  —
         * |*  |*|*              | * * *||  * |  * |  *   *|  |*  ** ||* |*   *
```

Fig. 2. (*Contd.*)

```
              1970      1980      1990      2000      2010      2020      2030
ToxinB  EAIDFTGK-LIIDENIYYFDDNYRGAVEWKELDGEMHYFSPETGKAFKGLNQIGDYKYYFNSD-GVMQKG
LT      ESVNFIGK-LNIDGKIYYFEDNYRAAVEWKLLDDETYYFNPKTGEALKGLHQIGDNKYYFDDN-GIMQTG
ToxinA  QAIVYQSKFLTLNGKKYYFDNNSKAVTGWRIINNEKYYFNPNNAIAAVGLQVIDNNKYYFNPDTAIISKG
ALPHA   RAINYTGW-LTLDGNKYYFQSNSKAVTGLQKISDKYYFN-DNGQMQIKWQIINNNKYYFDGNTGEAIIG

              2040      2050      2060      2070      2080      2090      2100
ToxinB  FVSINDNKHYFDDSGVMKV-GYTEIDGKHFYFAEN--------------GEMQIGVFNTEDGFKYFAHHNEDLGN
LT      FITINDKVFYFNNDGVMQV-GYIEVNGKYFYFGKN--------------GERQLGVFNTPDGFKFFGPKDDDLGT
ToxinA  WQTVNGSRYYFDTDTAIAFNGYKTIDGKHFYFDSD-------------CVVKIGVFSTSNGFEYFAPANTYNNN
ALPHA   WFNNNKERYYFDSEGRLLT-GYQVIGDKSYFSDNINGNWEEGSGVLKSGIFKTPSGFKLFSSEGD---K

              2110      2120      2130      2140      2150      2160      2170
ToxinB  EEGEEISY-SGILNFNNKIYYFDDSFTAVVG--------------WKDLE-------
LT      EEGELTLY-NGILNFNGKIYFFDISNTAVVG--------------WGTLD-------
ToxinA  IEGQAIVYQSKFLTLNGKKYYFDNNSKAVTGLQTIDSKKYYFNTNTAEAATGWQTIDGKKYYFNTNTAEA
ALPHA   ---SAINY-KGWLDLNGNKYYFNSDSIAVTG--------------SYN-------

              2180      2190      2200      2210      2220      2230      2240
ToxinB  -----DGSKYYFDEDTAEAYIGLSLINDGQYYFNDDGIMQVGFV-------
LT      -----DGSTYYFDDNRAEACIGLTVINDCKYYFDDNGIRQLGFI-------
ToxinA  ATGWQTIDGKKYYFNTNTAIASTGYTIINGKHFYFNTDGIMQIGVFKGPNGFEYFAPANTDANNIEGQAI
ALPHA   -----IKGIQYYFNPKTAVLTNG-------WY-------
```

Fig. 2. (Contd.)

```
               2250      2260      2270      2280      2290      2300      2310
ToxinB  ------------------------------------------------------------TINDKVFYFSDSGIIESGVQNI
LT      ------------------------------------------------------------TINDNIFYFSESGKIELGYQNI
ToxinA  LYQNEFLTLNGKKYYFGSDSKAVTGWRIINNKKYYFNPNNAIAAIHLCTINNDKYYFSYDGILQNGYITI
ALPHA   -----------------------------------------------------------TLDNNNYYVS-NGHNVLGYQDI
                                                                   |***  *

               2320      2330      2340      2350      2360      2370      2380
ToxinB  DDNYFYIDDNGIVQI--GVFDTSDGYKYFAPANTVNDNIYGQAVEYSG-LVRVGEDVYFGETYTIETGW
LT      NGNYFYIDESGLVLI--GVFDTPDGYKYFAPLNTVNDNIYGQAVKYSG-LVRVNEDVYFGETYKIETGW
ToxinA  ERNNFYFDANNESKMVTGVFKGPNGFEYFAPANTHNNNIEGQAIVYQNKFLTLNGKKYYFDNDSKAVTGW
ALPHA   DGKGYYFDPSTGIQKA-GVFPTPNGLRYFT-MKPIDGQRWGQCIDYTG-WLHLNGNKYYFGYYNSAVTGW
        * * *|*|   |||    *    |* * *  ||  *            *            |||

               2390      2400      2410      2420      2430      2440      2450
ToxinB  IYD--------------MENESDKYYFNPETKKACKGINLIDDIKYYFDEKG-IMRTGLISFENNN
LT      I---------------ENETDKYYFDPETKKAYKGINVDDIKYYFDENG-IMRTGLISFENNN
ToxinA  QTIDGKKYYFNLNTAEAATGWQTIDGKKYYFNLNTAEAATGWQTIDGKKYYFNTNTFIASTGYTSINGKH
ALPHA   R---------------VLGGKRYFFNIKTGAATTGLLTLSGKRYYFNEKG--EQLTLV------
            *|*|* *|     *  |||  *

               2460      2470      2480      2490      2500      2510      2520
ToxinB  YYFNENGEMQFG-----------------------------------------------------------
LT      YYFNEDGKMQFG-----------------------------------------------------------
ToxinA  FYFNTDGIMQIGVFKGPNGFEYFAPANTDANNIEGQAILYQNKFLTLNGKKYYFGSDSKAVTGLRTIDGK
ALPHA   -----------------------------------------------------------------------
```

Fig. 2. (*Contd.*)

```
              2530      2540      2550      2560      2570      2580      2590
              |         |         |         |         |         |         |
ToxinB        ------------------------------------------------------------
LT            ------------------------------------------------------------
ToxinA        -----------------------YINIEDKMFYFGEDGVMQIGVFNTPDGFKYFAHQN
                                     -YLNIKDKMFYFGKDGKMQIGVFNTPDGFKYFAHQN
ALPHA         KYYFNTNTAVAVTGWQTINGKKYYFNTNTSIASTGYTIISGKHFYFNTDGIMQIGVFKGPDGFEYFAPAN


              2600      2610      2620      2630      2640      2650      2660
              |         |         |         |         |         |         |
ToxinB        TLDENFEGESINY-----------------------------------------------
LT            TLDENFEGESINY-----------------------------------------------
ToxinA        TDANNIEGQAIRYQNRFLYLHDNIYYFGNNSKAATGWVTIDGNRYYFEPNTAMGANGYKTIDNKNFYFRN
              --------------------------TGWLDLDEKRYYFTD-------------------
              --------------------------TGWLDLDGKRYYFTD-------------------
ALPHA


              2670      2680      2690      2700      2710      2720      2730
              |         |         |         |         |         |         |
ToxinB        ------------EYIA--------------------------------------------
LT            ------------EYIA--------------------------------------------
ToxinA        GLPQIGVFKGSNGFEYFAPANTDANNIEGQAIRYQNRFLHLLGKIYYFGNNSKAVTGWQTINGKVYFMP
              -----------------------------------------ATGSVIIDGEEYYFDP
              -----------------------------------------ATGSLTIDGYNYFDP
ALPHA


              2740      2750      2760
              |         |         |
ToxinB        DTAQLVISE-----------------
LT            DTAELVVSE-----------------
ToxinA        DTAMAAAGGLFEIDGVIYFFGVDGVKAPGIYG
ALPHA
```

1 ToxinB   2366 aa Genbank Acc# X53138
2 LT   2364 aa Genbank Acc# X82638
3 ToxinA   2710 aa Genbank Acc# X51797
4 ALPHA   2178 aa Genbank Acc# Z48636

Fig. 2. (*Contd.*)

for toxin A and B, the maximal binding capacity for each toxin, however, is different (CHAVES-OLARTE et al. 1997). This difference and the different binding to special carbohydrates indicate that the cellular binding sites are not identical (TUCKER and WILKINS 1991). Essential elements for binding of toxin A to the cell membrane are carbohydrate structures. Treatment of cells with glycosidases or the N-glycosylation inhibitor tunicamycin reduces toxin A effects on cells (POTHOULAKIS et al. 1996, 1991; SMITH et al. 1997). Furthermore, direct binding of toxin A to the terminal carbohydrate structure Galα1-3Galβ1-4GlcNAc was shown (KRIVAN et al. 1986; TUCKER and WILKINS 1991). This structure, however, is absent in humans (LARSEN et al. 1990). Recently, it was demonstrated that toxin A also binds to GalNAcβ1-3Galβ1-4GlcNAc which is present in humans (KARLSSON 1995). Binding of toxin A to cells and cell membranes was decreased by protease treatment, indicating the involvement of proteinaceous structures (POTHOULAKIS et al. 1991). Thus, the cell receptor of toxin A is most likely a glycoprotein. If the toxins have the properties of lectins, it cannot be excluded that the toxins bind to several structurally related carbohydrates in a saturable manner, but only one is the specific receptor which mediates the endocytosis of the toxins. This notion is supported by the findings that toxin A binds to carbohydrate structures which do not exist in humans and to immunoglobulin and non-immunoglobulin components of milk (ROLFE and SONG 1995).

More recently, the membranous sucrase-isomaltase glycoprotein was identified as toxin A receptor in rabbit ileal brush border (POTHOULAKIS et al. 1996). However, this receptor does not seem to be of general importance because it is not expressed in many toxin-sensitive tissues (e.g. human colon) (POTHOULAKIS et al. 1996). The repetitive structure of the receptor domains of toxin A and toxin B argues for a modular organisation of this domain. By this modular structure, the toxin can bind to several possibly identical receptors to induce clustering followed by endocytosis. The current concept of toxin A cell entry is that the toxin binds to its cellular receptor via a lectin-like binding. The receptor of toxin B has not been characterised yet.

After binding to the cell receptor, internalisation of toxin A takes place via clathrin-coated pits as has been shown by electron microscopy (KUSHNARYOV and SEDMARK 1989). Alkalisation of acidic endosomal compartments inhibits the cytotoxic effects which can be circumvented by short-term acidification of the extracellular medium. These findings are consistent with cell entry through receptor-mediated endocytosis and toxin release from acidic endosomes (FLORIN and THELESTAM 1983; HENRIQUES et al. 1987; FIORENTINI and THELESTAM 1991). The translocation step is proposed to take place analogous to the cell entry of the well-characterised diphtheria toxin (KAUL et al. 1996). The low pH in the endosomes induces refolding of the toxin resulting in the insertion of the transmembrane domain into the endosomal membrane followed by translocation of the catalytic domain. The intact cytotoxin molecule is necessary for cell entry but it is unknown whether the complete peptide chain or a proteolytically cleaved N-terminal fragment is translocated into the cytosol. Both the N-terminal catalytic domain and the holotoxins are cytotoxic after microinjection indicating that processing is not essential for intracellular activity.

Toxin A is believed to be the enterotoxin which primarily disrupts the colonic epithelium to allow toxin B to act on non-epithelial tissue compartments. This notion is based on the findings that only toxin A is cytotoxic in animal models and in animal intestinal explants. Colonic epithelial cells from animals do not seem to possess a receptor for toxin B. However, recent findings suggest that human colonic epithelial cells are equally or even more sensitive to toxin B than to toxin A. These findings show the significance of toxin B in human disease (HECHT et al. 1992; CHAVES-OLARTE et al. 1997; RIEGLER et al. 1995).

4 Molecular Mechanism of Cytotoxicity

4.1 Enzyme Activity

The *Clostridium difficile* toxins A and B induce morphological and cytoskeletal alterations which resemble the cytotoxic features caused by *Clostridium botulinum* exoenzyme C3. C3 is a 25kDa single chain mono-ADP-ribosyltransferase which selectively modifies the low molecular mass GTP-binding RhoA, B and C proteins. The ADP-ribose moiety is linked to Asn-41 resulting in deactivation of the GTP-binding proteins (AKTORIES and KOCH 1997). Using C3-catalysed ADP-ribosylation as a probe for cellular Rho, it was found that intoxication of intact cells with toxin A or B resulted in an inhibition of ADP-ribosylation of Rho (JUST et al. 1994, 1995a). All findings with C3 as a tool indicated a stable modification of Rho which prevented ADP-ribosylation. Mass spectrometric analysis of toxin-modified Rho revealed an increase in mass of 162Da, consistent with the incorporation of a hexose. Biochemical studies identified the cofactor as UDP-glucose and the modification as glucosylation, i.e. incorporation of one glucose moiety (JUST et al. 1995b,c; AKTORIES and JUST 1995).

Toxin A and B are mono-glucosyltransferases which recruit UDP-glucose as cosubstrate to transfer the glucose moiety to the substrate protein (Fig. 3). The acceptor amino acid is threonine (Thr-37 in RhoA) consistent with an O-glucosylation type of reaction.

4.1.1 Cosubstrates

Toxin A (JUST et al. 1995b), toxin B (JUST et al. 1995c), lethal (JUST et al. 1996; POPOFF et al. 1996) and haemorrhagic toxin (GENTH et al. 1996) use UDP-glucose whereas α-toxin (SELZER et al. 1996) recruits UDP-N-acetyl-glucosamine (UDP-GlcNAc) as cosubstrate (Table 1). The activated nucleotide sugar is cleaved into glucose/GlcNAc and UDP and the sugar moiety is transferred to the acceptor amino acid of the Rho proteins. Glucose/GlcNAc is bound to the hydroxyl group of the acceptor residue threonine (JUST et al. 1995b,c; POPOFF et al. 1996). The incorporation does not exceed one molecule of glucose per molecule of Rho pro-

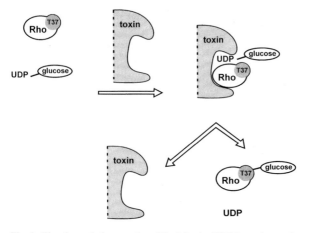

Fig. 3. The glucosylation reaction. RhoA in the GDP-bound state forms a ternary complex with toxin B and the cosubstrate UDP-glucose. The nucleotide sugar is cleaved and glucose moiety is transferred to threonine-37 (*T37*) of RhoA. The complex dissociates and releases glucosylated RhoA and UDP. Toxin B is ready to start a new glucosylation reaction. The glucosylation is in principle a reversible reaction: in the absence of UDP-glucose, but in an excess of UDP, toxin B deglucosylates RhoA to form UDP-glucose

tein, a finding which is in agreement with a monoglucosylation reaction (JUST et al. 1995b,c).

Only UDP-glucose and UDP-*N*-acetyl-glucosamine are recruited as cosubstrates, and not other nucleotide sugars (JUST et al. 1995b,c). The intracellular UDP-glucose concentration is estimated at 100μM (LAUGHLIN et al. 1988) indicating that the cosubstrate is not a limiting factor for the intracellular enzyme activity of the toxins. That UDP-glucose is the exclusive cosubstrate for toxin A and B also in intact cells has been demonstrated in a mutant Don cell line which is deficient of UDP-glucose. These cells are resistant to toxin A and B but the resistance can be overcome by microinjection with UDP-glucose (CHAVES-OLARTE et al. 1996).

There is one indirect evidence for in vivo glucosyltransferase activity of toxin A and B within the disease. A positron emission tomography (PET) of a patient with *Clostridium difficile*-associated diarrhoea showed an abnormal uptake of [¹⁸F] fluorodeoxyglucose into the colonic wall, which was interpreted as an uptake in inflamed tissue (HANNAH et al. 1996). However, this finding could also be explained as metabolism of [¹⁸F]fluorodeoxyglucose to the UDP conjugate, which is then cosubstrate for toxin A and B.

4.1.2   Protein Substrates

4.1.2.1   Rho and Ras Proteins as Substrates

The substrate specificity of the cytotoxins has been determined by testing recombinant low molecular mass GTPases for the property to be glucosylated in the presence of radioactive labelled UDP-glucose or UDP-*N*-acetyl-glucosamine. Toxin A and toxin B glucosylate RhoA, Rac1 and Cdc42 whereas other prototypes of the superfamily of low molecular mass GTPase such as H-Ras, Rab5 and Arf1 are not substrates (JUST et al. 1995b,c). Both toxins seem to possess the identical protein substrate specificity. However, recently it was reported that toxin A but not toxin B modifies Rap2 to a minor extent, indicating a real difference in the substrate specificity of the *C. difficile* cytotoxins (CHAVES-OLARTE et al. 1997) (see Table 1). Nevertheless, α-toxin from *Clostridium novyi,* which catalyses the incorporation of *N*-acetyl-glucosamine instead of glucose, modifies the same targets as toxin B, namely Rho, Rac and Cdc42 (SELZER et al. 1996).

The substrate specificity of lethal toxin, however, is not restricted to one GTPase subfamily but covers the Rho and Ras subfamily. Furthermore, the specificity depends on the strain from which lethal toxin is purified. Table 2 gives a synopsis of the substrates of the different isoforms of lethal toxin known so far (JUST et al. 1996; HOFMANN et al. 1996; POPOFF et al. 1996). Toxin B from *Clostridium difficile* strain 1470 resembles the lethal toxin and is therefore subsumed here (SCHMIDT et al. 1998).

Rho and Ras proteins are the preferred substrates for the large clostridial cytotoxins. So far, the substrates, with the exception of RhoA, were found by testing recombinant GTPases. In cell lysates the cytotoxins catalyse incorporation of [^{14}C]glucose and [^{14}C]*N*-acetyl-glucosamine, which is easily detected by phosphorimaging. The cellular concentration of Rho is estimated at about 1μM (PRICE et al. 1995), whereas other low molecular GTPases such as Ras are less concentrated in cells and detection of [^{14}C]glucosylated proteins is less sensitive or impossible without previous concentration. The diverse substrate specificity of the isoforms of lethal toxin clearly shows that testing one single prototype of GTPase subfamily does not exclude the other members of the subfamily as substrates. Thus, it is conceivable that the toxins modify more substrates than identified so far.

4.1.2.2   Site of Modification

Glucose is transferred by the cytotoxins to threonine-37 in Rho. In Rac, Cdc42 and Ras it is threonine-35, which is equivalent to threonine-37 in Rho. The acceptor amino acids were determined by sequencing (JUST et al. 1995b,c) and the identified threonine-37 in Rho was corroborated by mutation (JUST et al. 1995c; SELZER et al.

Table 2. Protein substrate specificity of lethal toxin from various *Clostridium sordellii* strains

C. sordellii strain	Protein substrates						Reference
CN6	(Rho)	Rac	(Cdc42)	Ras	Rap	Ral	I. Just, unpublished data
VPI9048	–	Rac	(Cdc42)	Ras	Rap	–	HOFMANN et al. 1996
6018	–	Rac	–	Ras	Rap	Ral	JUST et al. 1996
IP82	–	Rac	–	Ras	Rap	–	POPOFF et al. 1996
C.d. B-1470	–	Rac	–	–	(Rap)	Ral	SCHMIDT et al. 1998

C.d. B-1470, *Clostridium difficile* toxin B-1470 resembles *Clostridium sordellii*; ( ), minor substrate.

1996). All large clostridial cytotoxins catalyse the modification of identical acceptor threonine (Table 1).

Threonine-37/35 is located in the effector domain of the low molecular mass GTPases where coupling with the effector proteins takes place resulting in downstream signalling. In addition, threonine-37/35 is involved in the co-ordination of the magnesium ion which binds to β- and γ-phosphates of GTP. Exclusively in the GTP-bound state, this threonine participates in GTP binding. After hydrolysis and release of the γ-phosphate, loop L2 moves and the threonine-37/35 exposes its hydroxyl group to the surface of the protein (PAI et al. 1989, 1990; WITTINGHOFER et al. 1993). Thus, the hydroxyl group is accessible for glucosylation only in the GDP-bound form. Experimental data support this consideration: GDP-bound Ras is monoglucosylated, i.e. one mol of glucose is incorporated into one mol of Ras; Ras bound to GTPγS, the non-hydrolysable GTP analogue, is not a substrate (HERRMANN et al. 1998). The same was found for N-acetyl-glucosaminylation of Cdc42 by α-toxin (SELZER et al. 1996). Thus, the recognition of Rho and Ras proteins by the cytotoxins strictly depends on a defined conformational state and is not merely determined by the primary or secondary structure of the substrate.

4.1.3 Catalysis

The three-domain structure of the large clostridial toxins has been deduced from sequence analysis (Fig. 1). Biological activity has been assigned to the N-terminal fragment. Direct proof for this assignment has been obtained from the separate expression of the toxin fragments and their testing for glucosyltransferase activity. Only the N-terminal fragment of toxin B covering amino acids 1 through 900 showed glucosyltransferase activity, whereas the intermediary part (aa 901–1750) and the C-terminal fragment (aa 1751–2366) were absolutely devoid of transferase activity (HOFMANN et al. 1997). The same was found for toxin A, where the enzymatic activity resides in the first 659 N-terminal amino acids (FAUST et al. 1998). Successive C-terminal truncation of the N-terminal fragment of toxin B led to a minimum fragment covering amino acids 1–546 of toxin B which exhibits full catalytic activity and is cytotoxic when microinjected into intact cells (HOFMANN et al. 1997). Further deletion resulted in an enzymatically inactive fragment (aa 1–516) (HOFMANN et al. 1997).

The active fragment showed binding to the cosubstrate, detected by labelling with azido-UDP-glucose, whereas the inactive fragment did not bind UDP-glucose (BUSCH et al. 1998). The putative nucleotide-binding site (aa 651–683) (BARROSO et al. 1994; LYERLY and WILKINS 1995) is clearly beyond the minimum catalytic fragment (aa 1–546) and is obviously not involved in the binding of UDP-glucose and the enzymatic activity.

Toxin B is reported to show about 100-fold higher enzyme activity than toxin A (CHAVES-OLARTE et al. 1997), with the reservation that this paper does not present exact enzyme kinetics. This difference in activity between both toxins was also found in intact cells. When the toxins were microinjected to circumvent the cell entry mechanism, the cytotoxic potency differed also by a factor of 100. It seems that the overall observed difference in cytotoxic potency of toxin A and B on intact cells can be explained by the mere difference in enzyme activity.

In addition to glucosyltransferase activity, the toxins exhibit glycohydrolase activity to hydrolytically cleave UDP-glucose (toxin A and B, lethal toxin) and UDP-GlcNAc (α-toxin) in the absence of protein substrates into UDP and glucose/ GlcNAc, respectively (JUST et al. 1995c; CHAVES-OLARTE et al. 1997; CIESLA and BOBAK 1998). The nucleotide specificity is identical with the transferase reaction, only UDP-glucose is hydrolysed (JUST et al. 1995c). The glycohydrolase activity is about one to two orders of magnitude less than the transferase activity. The biological relevance of the glycohydrolase activity is unclear.

4.1.3.1 Cofactors

Because the glycohydrolase activity is a pure two component system – consisting exclusively of UDP-glucose and toxin – and thus devoid of other interfering factors, it is suitable to study the conditions of the catalytic activity. Glycohydrolase activity is stimulated by K^+ but inhibited by Na^+ (CIESLA and BOBAK 1998). Thus, the hydrolase activity relies on specific effects of K^+ and not mere ionic strength (CIESLA and BOBAK 1998). In view of the intracellular activity of toxin A and B, K^+ regulation of the enzyme activity makes sense. In addition to the monovalent cation K^+, glycohydrolase activity is strongly stimulated by Mn^{2+} and less by Mg^{2+}. The Mn^{2+} dependence has also been reported for the glucosyltransferase activity of lethal toxin which shows about 90% homology to toxin B (JUST et al. 1996). The divalent cation Mn^{2+} is an essential cofactor for several UDP-glucose hydrolases as well as many glycosyltransferases (MCEUEN 1992). These transferases possess a conserved DXD (aspartic acid–any amino acid–aspartic acid) motif surrounded by a hydrophobic region (Fig. 4) (WIGGINS and MUNRO 1998). DXD is very likely to

```
                          DXD motif
                            ↓↓
Toxin A   249 ELLNIYSQELLNRGNLAAASDIVRLLALKNFGGVYLDVDMLPGIHSDLFKTISRPSSI 306

Toxin B   250 ESFNLYEQELVERWNLAAASDILRISALKEIGGMYLDVDMLPGIQPDLFESIEKPSSV 307

α-Toxin   248 KLKSYYYQELIQTNNLAAASDILRIAILKKYGGVYCDLDFLPGVNLSLFNDISKPNGM 305

LT        250 DLVRLYNQELVERWNLAAASDILRISMLKEDGGVYLDVDILPGIQPDLFKSINKPDSI 307

Och1p     151 APVPIVIQAFKLMPGNILKADFLRYLLLFARGGIYSDMDTMLLKPIDSWPSQNKSWLN 208

Sur1p     102 EEYPWFLDTFENYKYPIERADAIRYFILSHYGGVYIDLDDGCERKLDPLLAFPAFLRK 159

Mnn1p     394 LDVSNTIHPKWRGDFGSYKSKWLVVLLNLLQEFVFLDIDAISYEKIDNYFKTTEYQKT 451
```

Fig. 4. Alignment of the large clostridial cytotoxins and glycosyltransferases covering the DXD motif. *Clostridium difficile* toxin A and toxin B, *Clostridium novyi* α-toxin, *Clostridium sordellii* lethal toxin (*LT*), *Saccharomyces cerevisiae* Och1p mannosyltransferase, *S. cerevisiae* Sur1p mannosyltransferase and *S. cerevisiae* Mnn1p mannosyltransferase

be involved in manganese binding (WIGGINS and MUNRO 1998). This motif is also located in the N-terminal catalytic part of the large clostridial cytotoxins. Mutations of these conserved aspartic acids in lethal toxin (Asp286 and Asp288) completely abolish both glucosyltransferase and glycohydrolase activity (BUSCH et al. 1998). Furthermore, labelling of the N-terminal fragment of lethal toxin with azido-UDP-glucose is blocked in the DXD deficient fragment even in the presence of high concentrations of Mn^{2+} (BUSCH et al. 1998). Thus, this motif represents a structure essential for catalysis. The aspartic acids are thought to participate in the co-ordination of Mn^{2+}, and Mn^{2+} allows correct positioning of the cosubstrate UDP-glucose which is subsequently catalytically cleaved and transferred to acceptor amino acid threonine. The catalytic amino acid of the cytotoxins has not been identified yet.

4.1.3.2 Recognition of the Protein Substrates

For glucosylation, the toxin has to form a ternary complex with UDP-glucose and the substrate protein (Rho proteins). Therefore, the toxins need, in addition to the UDP-glucose binding site and catalytic centre, a substrate recognition site. Whereas toxin B exclusively modifies the Rho subfamily proteins, lethal toxin glucosylates Rac and Cdc42 from the Rho subfamily and Ras, Ral and Rap from the Ras subfamily. It is unlikely that there is only one single substrate recognition site which is able to recognise variant combinations of the Rho and Ras GTP-binding proteins. Rather, it is conceivable that the toxins possess different recognition sites. This notion explains the differences in substrate specificity of the isoforms of lethal toxin and the variant toxin of *Clostridium difficile* (Table 2).

Because toxin B and lethal toxin are highly homologous (90%) but differ in their protein substrate specificity, chimeric toxins are helpful to restrict the site of substrate recognition. Testing the substrate specificity of various chimeras between the N-terminal part of toxin B and lethal toxin led to the restriction and separation of recognition sites for the Rho and Ras proteins. Amino acids 408–468 in toxin B determine the specificity for Rho, Rac and Cdc42, whereas residues 364–408 in lethal toxin determine for Rac and Cdc42 but not for Rho. Interestingly, lethal toxin and toxin B recruit different domains to be specific for Rho proteins. The recognition of Ras proteins is mediated from the region aa 408–516 which is adjacent to the Rho recognition site (HOFMANN et al. 1998). It seems that the substrate recognition sites are modularly organised.

4.2 Functional Consequences of Glucosylation

4.2.1 Cellular Functions of the Rho Proteins

Rho proteins belong to the Ras superfamily of low molecular mass GTP-binding proteins (GTPases) which are involved in the intracellular signal transduction to serve as molecular switches. These GTPases are characterised by their molecular weight (18–28kDa), their C-terminal polyisoprenylation and their property to bind and hydrolyse guanine nucleotides. They are inactive in the GDP-bound state and binding of GTP induces activation resulting in downstream signalling. The transition between inactive and active state is controlled by several regulatory proteins: guanine nucleotide exchange factors (GEF) promote the exchange of nucleotides and binding of GTP; GTPase-activating proteins (GAP) strongly stimulate the low intrinsic GTPase activity to terminate the activated state; guanine nucleotide dissociation inhibitor (GDI) traps the inactive GDP-bound form in a high affinity complex. Binding of GTP induces changes in the conformation which allows binding to the so-called effector protein. Effectors are often serine/threonine kinases which possess a Rho-binding domain. Binding of Rho results in activation of the kinase (e.g. ROKα/Rho kinase) which phosphorylates downstream targets. In addition to kinases, Rho effectors also comprise multidomain proteins without enzymatic activity (rhotekin and rhophilin) which may serve as a nucleus for multiprotein complexes to connect different signalling pathways. The signalling cycle of the Rho proteins is depicted in Fig. 5 (reviews on the Rho proteins, MACHESKY and HALL 1996; NARUMIYA 1996; RIDLEY 1996; TAPON and HALL 1997; VAN AELST and D'SOUZA-SCHOREY 1997; HALL 1998; MACKAY and HALL 1998; SASAKI and TAKAI 1998; AMANO et al. 1998).

The best characterised members of the Rho subfamily are Rho, Rac and Cdc42. In general, they are involved in the regulation of the dynamic actin cytoskeleton. Each of them, however, regulates distinct structures: Rho governs the formation of stress fibres and focal adhesions, Rac is involved in membrane ruffling and Cdc42 in the formation of filopodia. The best understood functional module is the formation of the stress fibres: Rac and Rho regulate the phosphatidylinositide (4)-phosphate-5-kinase (PI-5-kinase) to form PIP_2. PIP_2 (phosphatidylinositol 4,5-biphosphate)

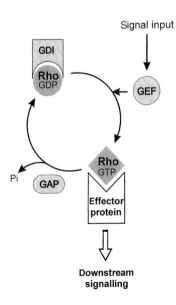

Fig. 5. GTPase cycle of Rho. Inactive Rho is GDP-bound and complexed with *GDI* (guanine-nucleotide dissociation inhibitor). Activation of Rho is mediated by *GEF* (guanine-nucleotide exchange factor) which promotes binding of GTP to Rho and translocation from the cytosolic to the membrane compartment. Here, Rho-GTP interacts with effector proteins (serine/threonine kinases or adapter proteins) which transmit the signal downstream. The active state is terminated by *GAP* (GTPase activating proteins) which strongly increases the intrinsic GTPase activity resulting in inactive GDP-bound Rho. GDP-bound Rho is extracted by *GDIs* from the membrane and complexed as inactive Rho pool in the cytosol

stimulates actin polymerisation and filament growth through interaction with several actin-binding proteins (e.g. gelsolin and profilin). The stress fibres, a supra-organisation of actin filaments, are governed by the RhoA-dependent Rho-kinase which phosphorylates the myosin light chain thereby activating the actin–myosin system in non-muscle cells. The membrane attachment of the stress fibres is managed through the ERM proteins (ezrin, radixin and moesin). These bifunctional proteins bind through their N-terminal part to transmembrane proteins (CD44 or ICAM proteins) and interact through their C-terminal part with the actin filaments. This interaction is essential for Rho-governed cytoskeletal changes.

The Rho proteins are involved in several cellular events. Table 3 gives a short overview of their multiple functions: Rho/Rac/Cdc42-dependent signal pathways are stimulated by receptor-tyrosine kinases (PDGF for Rac) and by G-protein-coupled receptors (LPA receptor for Rho, bradykinin receptor for Cdc42). The regulation of the exchange factors (GEFs) which directly promote activation of the Rho proteins is quite unclear but there are some data that tyrosine phosphorylation or phospholipid (PIP_2) binding is involved. The functional hierarchy of Rho proteins (Cdc42 activates Rac and Rac activates Rho) and the crosstalk between the proto-oncogene Ras and Rac is also part of the upstream regulation of Rho proteins.

4.2.2 Effects of Glucosylation on the GTPase Cycle

The cellular functions of low molecular mass GTPases of the Rho and Ras subfamily are determined by their ability to cycle between the inactive and active state, a process which is regulated by various regulatory proteins. Because the Ras GTPase cycle is very well characterised, all aspects of the functional consequences of glucosylation were studied with recombinant proteins (HERRMANN et al. 1998): (a) the activation of Ras by exchange factors (GEFs) is decreased but not completely inhibited, (b) the coupling to the effector protein Raf-kinase is completely blocked, (c) intrinsic GTPase is reduced but the GAP-stimulated GTPase is completely inhibited, and (d) no regulatory protein of the Ras cycle is sequestered arguing against a dominant negative mode of action.

Table 3. Cellular functions of the Rho proteins

Involvement	Functions
Organisation of the actin cytoskeleton	Stress fibres, membrane ruffling, filopodia formation, cell adhesion, cell–cell contact, cell morphology, cell motility
Membrane trafficking	Endocytosis, exocytosis, phagocytosis
Smooth muscle contraction	Calcium sensitisation
Phospholipid metabolism	PI-5-kinase, PLD, PLC
Cell cycle progression	Transition from G1 to S phase
Reactive oxygen species (ROS)	NADPH oxidase of neutrophils
Transcriptional activation	JNK, p38, NFκB
Cell transformation	Co-operation with the proto-oncogene Ras
Apoptosis	

Comparable data were found for glucosylated RhoA except that interaction with exchange factors was not studied (SEHR et al. 1998). Thus, the crucial step of downstream signalling, effector coupling, is completely inhibited. Figure 6 shows a schematic model of the influence of glucosylation on the GTPase cycle of Ras.

4.2.3 Biological Consequences of Rho Glucosylation

The actin depolymerising activity of the cytotoxins can be explained by the inactivation of Rho. The serine/threonine kinase ROK which is stimulated by activated RhoA directly phosphorylates the myosin light chain (MLC) to enhance formation of stress fibres. Inactivation of the myosin light chain by the MLC phosphatase is blocked by ROK-catalysed phosphorylation of the myosin-binding protein (MBP). Decrease in ROK activity by inactivation of Rho results in an increased activity of phosphatases which finally leads to dephosphorylation of MLC and subsequent depolymerisation of stress fibres. The growth of actin filaments is regulated by actin-binding proteins whose activity is stimulated by PIP_2 formed by the Rho/Rac-regulated phospholipid kinase. Inactivation of Rho/Rac leads, through decrease in PIP_2 formation, to an impaired F-actin stability. The actin cytoskeleton is dynamically regulated by polymerising and depolymerising signals. Shut-down of the polymerising signals by the large clostridial cytotoxins leads to the predomi-

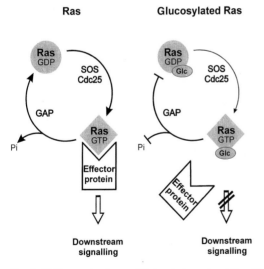

Fig. 6. GTPase cycle of unmodified and glucosylated Ras. The *left panel* gives the GTPase cycle of Ras. Inactive GDP-bound Ras is activated by exchange factors such as SOS or Cdc25. Ras-GTP binds to the effector proteins, e.g. Raf kinase, which are activated to transmit the signal downstream. The active state is terminated by *GAP* (GTPase activating proteins) which promotes inactivation of Ras. The *right panel* gives the cycle of glucosylated Ras. Ras in the GDP-bound form is monoglucosylated at threonine-35. Loaded with GTP catalysed by SOS or Cdc25, it is slowed down but not inhibited. Because glucosylated Ras is unable to interact with *GAP*, it is enriched in the GTP-bound form. However, interaction with effector proteins such as Raf kinase is fully blocked, resulting in a complete inhibition of downstream signalling

nance of the depolymerising inputs. One of these depolymerising inputs could be the RhoE/Rnd3 protein, a low molecular mass GTP-binding protein belonging to the Rho subfamily. RhoE/Rnd1, which is permanently bound to GTP because of lacking GTP hydrolysing activity, induces rounding of cells and loss of stress fibres (NOBES et al. 1998; GUASCH et al. 1998). This Rho subtype is not a substrate for toxin A or B (I. Just, unpublished data) and thus can functionally predominate after inactivation of Rho/Rac/Cdc42 by glucosylation.

The primary target tissue of *Clostridium difficile* toxins is the colonic epithelium. Human colon carcinoma (T84) cells are polarised in culture forming tight junctions and therefore they are an established model for the colonic epithelial barrier. Toxin A and B disrupt the barrier function by opening the tight junctions (MOORE et al. 1990; TRIADAFILOPOULOS et al. 1987; HECHT et al. 1988, 1992). This effect is not merely caused by the breakdown of actin filaments but by inactivation of the Rho function to regulate tight junction complex (NUSRAT et al. 1995; JOU et al. 1998). These barrier-disrupting effects of toxin A and B are supposed to increase the colonic permeability, the basis of the watery diarrhoea, which is a typical feature of the *Clostridium difficile*-induced antibiotic-associated diarrhoea.

Toxin A and B have been reported to induce apoptosis (MAHIDA et al. 1996; FIORENTINI et al. 1998; CALDERON et al. 1998) consistent with the function of Rho in the apoptotic process (GOMEZ et al. 1998). Apoptosis is induced by many signals and one is the detachment of cells from their extracellular matrix (RUOSLAHTI 1997). The cytotoxins, which are known to induce detachment through their actin filament-disrupting properties, induce apoptosis like the treatment of cells with EDTA or neutrophil elastase (SHIBATA et al. 1996).

Besides accessory effector domains, the domain covering amino acids 30–42 is thought to be the essential effector domain for downstream signalling, and a glucose moiety is likely to inhibit all signal pathways downstream of Rho and Ras. The cytotoxic effect is easy to observe and has led to their classification as cytotoxins. However, the depolymerisation of the actin filament system is only one of many cellular responses to the inactivation of Rho, Rac and Cdc42. For example, regulation of the endocytosis in *Xenopus* oocytes (SCHMALZING et al. 1995), the exocytosis of mediators from rat basophilic leukaemia (RBL) cells (PREPENS et al. 1996) or the regulation of phospholipase D (PLD) (SCHMIDT et al. 1998) are impaired by toxin B independently of the cytoskeletal effects.

Because glucosylation does not inhibit one single Rho function but many or even all, overlapping or unexpected findings after cell intoxication can be observed. For example, destruction of the actin cytoskeleton of RBL cells by cytochalasin D or the actin ADP-ribosylating *Clostridium botulinum* C2 toxin results in an increase of stimulated degranulation (exocytosis). However, intoxication with toxin B, which also leads to disruption of the actin filaments, inhibited stimulated degranulation – the opposite effect than was expected (PREPENS et al. 1996). Direct regulation of exocytosis by Rho is dominant over the influence of the actin filaments which are also governed by Rho. The large clostridial cytotoxins cannot be classified as simple cytoskeleton-disrupting toxins any longer since they interfere with many vital cell functions, especially those assigned to the Rho and Ras proteins.

5 Concluding Remarks

The large clostridial cytotoxins cause comparable morphological and cytoskeletal changes which are mediated by their intrinsic enzymatic activity to inactivate the regulators of the actin cytoskeleton, the Rho proteins. Comparable to other enzymatically active exotoxins, the large cytotoxins recruit ubiquitously available cofactors (nucleotide sugars) to alter pivotal regulators of the cell machinery. At the moment it is unclear why the toxins need 200–250kDa of protein to target their enzymatic domain (60kDa) to the cellular substrates. It may be that an additional biological activity is concealed in this part or the toxins have evolved a sophisticated targeting machinery which makes almost all cells sensitive to the toxins.

References

Aktories K, Just I (1995) Monoglucosylation of low-molecular-mass GTP-binding Rho proteins by clostridial cytotoxins. Trends Cell Biol 5:441–443

Aktories K, Koch G (1997) *Clostridium botulinum* ADP-ribosyltransferase C3. In: Aktories K (ed) Bacterial toxins: tools in cell biology and pharmacology. Chapman & Hall, Weinheim, pp 61–69

Amano M, Fukata Y, Kaibuchi K (1998) Regulation of cytoskeleton and cell adhesions by the small GTPase Rho and its targets. TCM 8:162–168

Barroso LA, Moncrief JS, Lyerly DM, Wilkins TD (1994) Mutagenesis of the *Clostridium difficile* toxin B gene and effect on cytotoxic activity. Microb Pathog 16:297–303

Bartlett JG (1994) *Clostridium difficile*: history of its role as an enteric pathogen and the current state of knowledge about the organism. Clin Infect Dis 18:265–272

Bette P, Oksche A, Mauler F, Von Eichel-Streiber C, Popoff MR, Habermann E (1991) A comparative biochemical, pharmacological and immunological study of *Clostridium novyi* α-toxin, *C. difficile* toxin B and *C. sordellii* lethal toxin. Toxicon 29:877–887

Busch C, Hofmann F, Selzer J, Munro J, Jeckel D, Aktories K (1998) A common motif of eukaryotic glycosyltransferases is essential for the enzyme activity of large clostridial cytotoxins. J Biol Chem 273:19566–19572

Calderón GM, Torres-López J, Lin T-J, Chavez B, Hernández M, Munoz O, Befus AD, Enciso JA (1998) Effects of toxin A from *Clostridium difficile* on mast cell activation and survival. Infect Immun 66:2755–2761

Chaves-Olarte E, Florin I, Boquet P, Popoff M, Von Eichel-Streiber C, Thelestam M (1996) UDP-glucose deficiency in a mutant cell line protects against glucosyltransferase toxins from *Clostridium difficile* and *Clostridium sordellii*. J Biol Chem 271:6925–6932

Chaves-Olarte E, Weidmann M, Von Eichel-Streiber C, Thelestam M (1997) Toxins A and B from *Clostridium difficile* differ with respect to enzymatic potencies, cellular substrate specificities, and surface binding to cultured cells. J Clin Invest 100:1734–1741

Ciesielski-Treska J, Ulrich G, Rihn B, Aunis D (1989) Mechanism of action of *Clostridium difficile* toxin B: role of external medium and cytoskeletal organization in intoxicated cells. Eur J Cell Biol 48: 191–202

Ciesielski-Treska J, Ulrich G, Baldacini O, Monteil H, Aunis D (1991) Phosphorylation of cellular proteins in response to treatment with *Clostridium difficile* toxin B and *Clostridium sordellii* toxin L. Eur J Cell Biol 56:68–78

Ciesla WP Jr, Bobak DA (1998) *Clostridium difficile* toxins A and B are cation-dependent UDP-glucose hydrolases with differing catalytic activities. J Biol Chem 273:16021–16026

Couet J, Li S, Okamoto T, Ikezu T, Lisanti MP (1997) Identification of peptide and protein ligands for the caveolin-scaffolding domain. J Biol Chem 272:6525–6533

Faust C, Ye B, Song K-P (1998) The enzymatic domain of *Clostridium difficile* toxin A is located within its N-terminal region. Biochem Biophys Res Commun 251:100–105

Fiorentini C, Arancia G, Paradisi S, Donelli G, Giuliano M, Piemonto F, Mastrantonio P (1989) Effects of *Clostridium difficile* toxins A and B on cytoskeleton organization in HEp-2 cells: a comparative morphological study. Toxicon 27:1209–1218

Fiorentini C, Malorni W, Paradisi S, Giuliano M, Mastrantonio P, Donelli G (1990) Interaction of *Clostridium difficile* toxin A with cultured cells: cytoskeletal changes and nuclear polarization. Infect Immun 58:2329–2336

Fiorentini C, Donelli G, Nicotera P, Thelestam M (1993) *Clostridium difficile* toxin A elicits Ca^{2+}-independent cytotoxic effects in cultured normal rat intestinal crypt cells. Infect Immun 61:3988–3993

Fiorentini C, Fabbri A, Falzano L, Fattorossi A, Matarrese P, Rivabene R, Donelli G (1998) *Clostridium difficile* toxin B induces apoptosis in intestinal cultured cells. Infect Immun 66:2660–2665

Fiorentini C, Thelestam M (1991) *Clostridium difficile* toxin A and its effects on cells. Toxicon 29:543–567

Florin I, Thelestam M (1983) Internalization of *Clostridium difficile* cytotoxin into cultured human lung fibroblasts. Biochim Biophys Acta 763:383–392

Frey SM, Wilkins TD (1992) Localization of two epitopes recognized by monoclonal antibody PCG-4 on *Clostridium difficile* toxin A. Infect Immun 60:2488–2492

Genth H, Hofmann F, Selzer J, Rex G, Aktories K, Just I (1996) Difference in protein substrate specificity between hemorrhagic toxin and lethal toxin from *Clostridium sordellii*. Biochem Biophys Res Commun 229:370–374

Gomez J, Martinez C, Gonzalez A, Rebollo A (1998) Dual role of Ras and Rho proteins – at the cutting edge of life and death. Immunol Cell Biol 76:125–134

Guasch RM, Scambler P, Jones GE, Ridley AJ (1998) RhoE regulates actin cytoskeleton organization and cell migration. Mol Cell Biol 18:4761–4771

Hall A (1998) Rho GTPases and the actin cytoskeleton. Science 279:509–514

Hannah A, Scott AM, Akhurst T, Berlangieri S, Bishop J, McKay WJ (1996) Abnormal colonic accumulation of fluorine-18-FDG in pseudomembranous colitis. J Nucl Med 37:1683–1685

Hecht G, Pothoulakis C, LaMont JT, Madara JL (1988) *Clostridium difficile* toxin A perturbs cytoskeletal structure and tight junction permeability of cultured human intestinal epithelial monolayers. J Clin Invest 82:1516–1524

Hecht G, Koutsouris A, Pothoulakis C, LaMont JT, Madara JL (1992) *Clostridium difficile* toxin B disrupts the barrier function of T_{84} monolayers. Gastroenterology 102:416–423

Henriques B, Florin I, Thelestam M (1987) Cellular internalisation of *Clostridium difficile* toxin A. Microb Pathogen 2:455–463

Herrmann C, Ahmadian MR, Hofmann F, Just I (1998) Functional consequences of monoglucosylation of H-Ras at effector domain amino acid threonine-35. J Biol Chem 273:16134–16139

Hofmann F, Rex G, Aktories K, Just I (1996) The Ras-related protein Ral is monoglucosylated by *Clostridium sordellii* lethal toxin. Biochem Biophys Res Commun 227:77–81

Hofmann F, Busch C, Prepens U, Just I, Aktories K (1997) Localization of the glucosyltransferase activity of *Clostridium difficile* toxin B to the N-terminal part of the holotoxin. J Biol Chem 272:11074–11078

Hofmann F, Busch C, Aktories K (1998) Chimeric clostridial cytotoxins: identification of the N-terminal region involved in protein substrate recognition. Infect Immun 66:1076–1081

Jou T-S, Schneeberger EE, Nelson WJ (1998) Structural and functional regulation of tight junctions by RhoA and Rac1 small GTPases. J Cell Biol 142:101–115

Just I, Richter H-P, Prepens U, Von Eichel-Streiber C, Aktories K (1994) Probing the action of *Clostridium difficile* toxin B in *Xenopus laevis* oocytes. J Cell Science 107:1653–1659

Just I, Selzer J, Von Eichel-Streiber C, Aktories K (1995a) The low molecular mass GTP-binding protein Rho is affected by toxin A from *Clostridium difficile*. J Clin Invest 95:1026–1031

Just I, Selzer J, Wilm M, Von Eichel-Streiber C, Mann M, Aktories K (1995b) Glucosylation of Rho proteins by *Clostridium difficile* toxin B. Nature 375:500–503

Just I, Wilm M, Selzer J, Rex G, Von Eichel-Streiber C, Mann M, Aktories K (1995c) The enterotoxin from *Clostridium difficile* (ToxA) monoglucosylates the Rho proteins. J Biol Chem 270:13932–13936

Just I, Selzer J, Hofmann F, Green GA, Aktories K (1996) Inactivation of Ras by *Clostridium sordellii* lethal toxin-catalyzed glucosylation. J Biol Chem 271:10149–10153

Karlsson KA (1995) Microbial recognition of target-cell glycoconjugates. Curr Opin Struct Biol 5: 622–635

Kaul P, Silverman J, Shen WH, Blanke SR, Huynh PD, Finkelstein A, Collier RJ (1996) Roles of Glu 349 and Asp 352 in membrane insertion and translocation by diphtheria toxin. Protein Sci 5:687–692

Kelly CP, Pothoulakis C, LaMont JT (1994) *Clostridium difficile* colitis. New England J Med 330(4): 257–262

Kelly CP, LaMont JT (1998) *Clostridium difficile* infection. Annu Rev Med 49:375–390

Kink JA, Williams JA (1998) Antibodies to recombinant *Clostridium difficile* toxins A and B are an effective treatment and prevent relapse of *C. difficile*-associated disease in a hamster model of infection. Infect Immun 66:2018–2025

Krivan HC, Clark GF, Smith DF, Wilkins TD (1986) Cell surface binding site for *Clostridium difficile* enterotoxin: evidence for a glycoconjugate containing the sequence Galα1-3Galβ1-4GlcNAc. Infect Immun 53:573–581

Kushnaryov VM, Sedmark JJ (1989) Effect of *Clostridium difficile* enterotoxin A on ultrastructure of Chinese hamster ovary cells. Infect Immun 57 (12):3914–3921

Larsen RD, Rivera-Marrero CA, Ernst LK, Cummings RD, Lowe JB (1990) Frameshift and nonsense mutations in a human genomic sequence homologous to a murine UDP-Gal:b-$_D$-Gal(1,4)-$_D$-GlcNAc a(1,3)-galactosyltransferase cDNA. J Biol Chem 265:7055–7061

Laughlin MR, Petit WA, Dizon JM, Shulman RG, Barrett EJ (1988) NMR measurements of in vivo myocardial glycogen metabolism. J Biol Chem 263:2285–2291

Lisanti MP, Scherer PE, Tang Z, Sargiacomo M (1994) Caveolae, caveolin and caveolin-rich membrane domains: a signalling hypothesis. Trends Cell Biol 4:231–235

Lyerly DM, Lockwood DE, Richardson SH, Wilkins TD (1982) Biological activities of toxins A and B of *Clostridium difficile*. Infect Immun 35:1147–1150

Lyerly DM, Saum KE, MacDonald DK, Wilkins TD (1985) Effects of *Clostridium difficile* toxins given intragastrically to animals. Infect Immun 47:349–352

Lyerly DM, Phelps CJ, Toth J, Wilkins TD (1986) Characterization of toxins A and B of *Clostridium difficile* with monoclonal antibodies. Infect Immun 54:70–76

Lyerly DM, Wilkins TD (1995) *Clostridium difficile*. In: Blaser MJ, Smith PD, Ravdin JI et al. Infections of the Gastrointestinal Tract. Raven Press Ltd., New York, pp 867–891

Machesky LM, Hall A (1996) Rho: a connection between membrane receptor signalling and the cyto-skeleton. Trends Cell Biol 6:304–310

Mackay DJG, Hall A (1998) Rho GTPases. J Biol Chem 273:20685–20688

Mahida YR, Makh S, Hyde S, Gray T, Borriello SP (1996) Effect of *Clostridium difficile* toxin A on human intestinal epithelial cells: Induction of interleukin 8 production and apoptosis after cell detachment. Gut 38:337–347

Malorni W, Paradisi S, Dupuis ML, Fiorentini C, Ramoni C (1991) Enhancement of cell-mediated cytotoxicity by *Clostridium difficile* toxin A: an in vitro study. Toxicon 29(4/5):417–428

McEuen AR (1992) Manganese metalloproteins and manganese-activated enzymes. Inorganic Biochemistry 3:314–343

Moore R, Pothoulakis C, LaMont JT, Carlson S, Madara JL (1990) *C. difficile* toxin A increases intestinal permeability and induces Cl⁻. Am J Physiol 259:G165–G172

Narumiya S (1996) The small GTPase Rho: Cellular functions and signal transduction. J Biochem (Tokyo) 120:215–228

Nobes CD, Lauritzen I, Mattei M-G, Paris S, Hall A (1998) A new member of the Rho family, Rnd1, promotes disassembly of actin filament structures and loss of cell adhesion. J Cell Biol 141:187–197

Nusrat A, Giry M, Turner JR, Colgan SP, Parkos CA, Carnes D, Lemichez E, Boquet P, Madara JL (1995) Rho protein regulates tight junctions and perijunctional actin organization in polarized epithelia. Proc Natl Acad Sci USA 92:10629–10633

Okamoto T, Schlegel A, Scherer PE, Lisanti MP (1998) Caveolins, a family of scaffolding proteins for organizing "preassembled signaling complexes" at the plasma membrane. J Biol Chem 273: 5419–5422

Oksche A, Nakov R, Habermann E (1992) Morphological and biochemical study of cytoskeletal changes in cultured cells after extracellular application of *Clostridium novyi* alpha-toxin. Infect Immun 60:3002–3006

Pai EF, Kabsch W, Krengel U, Holmes KC, John J, Wittinghofer A (1989) Structure of the guanine-nucleotide-binding domain of the Ha-ras oncogene product p21 in the triphosphate conformation. Nature 341:209–214

Pai EF, Krengel U, Petsko GA, Goody RS, Kabsch W, Wittinghofer A (1990) Refined crystal structure of the triphosphate conformation of H-ras p21 at 1.35A resolution: implications for the mechanism of GTP hydrolysis. EMBO J 9:2351–2359

Popoff MR (1987) Purification and characterization of *Clostridium sordellii* lethal toxin and cross-reactivity with *Clostridium difficile* cytotoxin. Infect Immun 55:35–43

Popoff MR, Chaves OE, Lemichez E, Von Eichel-Streiber C, Thelestam M, Chardin P, Cussac D, Chavrier P, Flatau G, Giry M, Gunzburg J, Boquet P (1996) Ras, Rap, and Rac small GTP-binding proteins are targets for *Clostridium sordellii* lethal toxin glucosylation. J Biol Chem 271:10217–10224

Pothoulakis C, LaMont JT, Eglow R, Gao N, Rubins JB, Theoharides TC, Dickey BF (1991) Characterizing of rabbit ileal receptors for *Clostridium difficile* toxin A. J Clin Invest 88:119–125

Pothoulakis C, Gilbert RJ, Cladaras C, Castagliuolo I, Semenza G, Hitti Y, Montcrief JS, Linevsky J, Kelly CP, Nikulasson S, Desai HP, Wilkins TD, LaMont JT (1996) Rabbit sucrase-isomaltase contains a functional intestinal receptor for *Clostridium difficile* toxin A. J Clin Invest 98:641–649

Prepens U, Just I, Von Eichel-Streiber C, Aktories K (1996) Inhibition of FcÎRI-mediated activation of rat basophilic leukemia cells by *Clostridium difficile* toxin B (monoglucosyltransferase). J Biol Chem 271:7324–7329

Price LS, Norman JC, Ridley AJ, Koffer A (1995) The small GTPases Rac and Rho as regulators of secretion in mast cells. Curr Biol 5:68–73

Ridley AJ (1996) Rho: theme and variations. Curr Biol 6:1256–1264

Riegler M, Sedivy R, Pothoulakis C, Hamilton G, Zacheri J, Bischof G, Cosentini E, Feil W, Schiessel R, LaMont JT, Wenzl E (1995) *Clostridium difficile* toxin B is more potent than toxin A in damaging human colonic epithelium in vitro. J Clin Invest 95:2004–2011

Rolfe RD, Song W (1995) Immunoglobulin and non-immunoglobulin components of human milk inhibit *Clostridium difficile* toxin A-receptor binding. J Med Microbiol 42:10–19

Ruoslahti E (1997) Stretching is good for a cell. Science 276:1345–1346

Sasaki T, Takai Y (1998) The Rho small G protein family-Rho GDI system as a temporal and spatial determinant for cytoskeletal control. Biochem Biophys Res Commun 245:641–645

Schmalzing G, Richter HP, Hansen A, Schwarz W, Just I, Aktories K (1995) Involvement of the GTP binding protein Rho in constitutive endocytosis in *Xenopus laevis* oocytes. J Cell Biol 130: 1319–1332

Schmidt M, Vo M, Thiel M, Bauer B, Grannass A, Tapp E, Cool RH, De Gunzburg J, Von Eichel-Streiber C, Jakobs KH (1998) Specific inhibition of phorbol ester-stimulated phospholipase D by *Clostridium sordellii* lethal toxin and *Clostridium difficile* toxin B-1470 in HEK-293 cells. J Biol Chem 273:7413–7422

Sehr P, Joseph G, Genth H, Just I, Pick E, Aktories K (1998) Glucosylation and ADP-ribosylation of Rho proteins-effects on nucleotide binding, GTPase activity, and effector-coupling. Biochemistry 37:5296–5304

Selzer J, Hofmann F, Rex G, Wilm M, Mann M, Just I, Aktories K (1996) *Clostridium novyi* a-toxin-catalyzed incorporation of GlcNAc into Rho subfamily proteins. J Biol Chem 271:25173–25177

Shibata Y, Nakamura H, Kato S, Tomoike H (1996) Cellular detachment and deformation induce IL-8 gene expression in human bronchial epithelial cells. J Immunol 156:772–777

Siffert J-C, Baldacini O, Kuhry J-G, Wachsmann D, Benabdelmoumene S, Faradji A, Monteil H, Poindron P (1993) Effects of *Clostridium difficile* toxin B on human monocytes and macrophages: possible relationship with cytoskeletal rearrangement. Infect Immun 61:1082–1090

Smith JA, Cooke DL, Hyde S, Borriello SP, Long RG (1997) *Clostridium difficile* toxin A binding to human intestinal epithelial cells. J Med Microbiol 46:953–958

Tapon N, Hall A (1997) Rho, Rac and CDC42 GTPases regulate the organization of the actin cytoskeleton. Curr Opin Cell Biol 9:86–92

Thompson JD, Higgins DG, Gibson TJ, Clustal W (1994) Improving the sensitivity of progressive multiple sequence alignment through sequence weighting, positions-specific gap penalties and weight matrix choice. Nucl Acids Res 22:4673–4680

Triadafilopoulos G, Pothoulakis C, O'Brien MJ, LaMont JT (1987) Differential effects of *Clostridium difficile* toxins A and B on rabbit ileum. Gastroenterology 93(2):273–279

Tucker KD, Wilkins TD (1991) Toxin A of *Clostridium difficile* binds to the human carbohydrate antigens I, X, and Y. Infect Immun 59:73–78

Van Aelst L, D'Souza-Schorey C (1997) Rho GTPases and signaling networks. Genes Dev 11:2295–2322

Von Eichel-Streiber C, Laufenberg-Feldmann R, Sartingen S, Schulze J, Sauerborn M (1992a) Comparative sequence analysis of the *Clostridium difficile* toxins A and B. Mol Gen Genet 233:260–268

Von Eichel-Streiber C, Sauerborn M, Kuramitsu HK (1992b) Evidence for a modular structure of the homologous repetitive C-terminal carbohydrate-binding sites of *Clostridium difficile* toxins and *Streptococcus mutans* glucosyltransferases. J Bacteriol 174:6707–6710

Von Eichel-Streiber C (1993) Molecular Biology of the *Clostridium difficile* Toxins. In: Sebald M (ed) Genetics and Molecular Biology of Anaerobic Bacteria. Springer-Verlag, New York, pp 264–289

Wiggins CAR, Munro S (1998) Activity of the yeast MNN1 a-1,3-mannosyltransferase requires a motif conserved in many other families of glycosyltransferases. Proc Natl Acad Sci USA 95:7945–7950

Wittinghofer A, Pai EF, Goody RS (1993) Structural and mechanistic aspects of the GTPase reaction of H-ras p21. In: Dickey F, Birnbaumer L (eds) GTPases in Biology I. Springer-Verlag, Berlin, Heidelberg, pp 195–211

Wren BW (1991) A family of clostridial and streptococcal ligand-binding proteins with conserved C-terminal repeat sequences. Mol Microbiol 5:797–803

Cytotoxic Effects of the *Clostridium difficile* Toxins

M. THELESTAM and E. CHAVES-OLARTE

1 Introduction

Clostridium difficile-induced antibiotic-associated diarrhea and pseudomembranous colitis are typical toxin diseases elicited by actions of the two major toxins A and B (TcdA, TcdB) in the intestine. TcdA and TcdB are cytotoxic to intestinal and other cells because they glucosylate small GTP-binding proteins. These GTPases are crucial proteins controlling the actin cytoskeleton (ACSK) and the molecular signaling pathways involved in cell proliferation and cell death. Besides the two "classic" toxins produced by most strains of *C. difficile*, certain strains produce variant toxins whose pathophysiological significance is still unclear (KATO et al. 1998; RUPNIK et al. 1998). The C. *difficile* toxins are prototypes of the family of so-called Large Clostridial cytoToxins (LCTs). The LCTs glucosylate a variety of small GTPases thereby inducing a collapse of the ACSK (EICHEL-STREIBER et al. 1996).

Microbiology and Tumorbiology Center, Karolinska Institutet, Box 280, 171 77 Stockholm, Sweden

In this chapter we will discuss the intoxication of cells exposed to the *C. difficile* toxins, including the entire sequence of events from cell surface binding of the toxins to the gross cellular consequences of GTPase glucosylations (summarised in Fig. 1). We will also highlight the current intensive efforts to understand cellular aspects of the pathogenesis of *C. difficile* disease.

2 Cytotoxicity

2.1 General Scheme for Cellular Intoxication

2.1.1 TcdA and TcdB Bind to Specific Receptors

Specific binding to receptors located on the cell surface is the first step in cellular intoxication mediated by the *C. difficile* toxins. TcdA binds to the trisaccharide Galα1-3Galβ1-4GlcNAc occurring on rabbit red blood cells and hamster brush border membranes (KRIVAN et al. 1986). In addition, TcdA binds to the Galβ1-4GlcNAc structure found on certain human blood group antigens (TUCKER and WILKINS 1991) and to GalNAcβ1-3Galβ1-4GlcNAc (TENEBERG et al. 1996), which is also present in human tissues. Thus, the disaccharide Galβ1-4GlcNAc appears to be the minimum structure required for binding of TcdA. Whether a membrane protein or lipid carries the carbohydrate responsible for the specific TcdA binding is not known. Lewis X, Y, and I antigens (TUCKER and WILKINS 1991) and sucrase-isomaltase (POTHOULAKIS et al. 1996) have been shown to bind TcdA, but the relevant receptor structure(s) in human colonic cells has not been determined yet.

The receptor structure for TcdB is currently not known. However, specific binding of TcdB to Don fibroblasts and to human colon carcinoma T84 cells was recently demonstrated (CHAVES-OLARTE et al. 1997). The specific binding of TcdB to these two cell lines correlated with their respective sensitivities to the toxin, indicating that the ability of TcdB to bind to its receptor is a limiting step in the cellular intoxication process. In addition, polarized T84 cells cultivated on permeable filters are more sensitive to TcdB applied at the basolateral side, suggesting that the relevant receptor for TcdB may be more concentrated on this side of the colonic epithelium (E. Chaves-Olarte et al., unpublished data).

2.1.2 TcdA and TcdB Enter Cells by Receptor-Mediated Endocytosis

After binding to their respective specific receptors on the cell surface, both toxins enter the cell by endocytosis. This process was visualized for TcdA by transmission electron microscopy, suggesting that the uptake occurs by receptor-mediated endocytosis (EICHEL-STREIBER et al. 1991). Both TcdA and TcdB were reported to require passage through an acidic intracellular compartment(s) in order to intoxicate cells (HENRIQUES et al. 1987). TcdB appeared to require delivery to lysosomes before its release to the cytosol (FLORIN and THELESTAM 1986) whereas the route of TcdA is not clear. Both TcdA and TcdB are able to intoxicate cells upon micro-injection, indicating that they can act intracellularly, without specific intravesicular processing (MÜLLER et al. 1992; CHAVES et al. 1997). Furthermore, the holotoxins had the same enzymatic potency in vitro as the catalytic N-terminal fragments (HOFMANN et al. 1997). Thus, if intracellular processing (of the extracellularly added and endocytosed toxins) takes place, this is not needed for activation. Nevertheless, cleavage of the toxins may still be a requisite for release of the cat-alytic fragments from the respective endocytic compartments into the cytosol. The nature of this putative enzymatic processing and the molecular details of toxin membrane translocation are not known.

2.1.3 TcdA and TcdB are Glucosyltransferases

Once the catalytic fragments of TcdA and TcdB reach the cytosol of the intoxicated cell they glucosylate members of the family of small GTP-binding proteins (JUST et al. 1995a,b). Both toxins modify and thus inactivate Rho, Rac, and Cdc42. Additionally, TcdA inactivates Rap 1 and 2 (CHAVES-OLARTE et al. 1997). The modification involves the transfer of a glucose moiety from UDP-glucose (UDP-Glc) to a conserved threonine located in the effector region of the target GTPase. This covalent modification impairs proper interaction of the targets with their downstream effector molecules, thereby interrupting the signaling pathways me-diated by these small GTPases. In vitro experiments with recombinant targets indicate that they are most efficiently modified when loaded with GDP. In this conformation, the lateral chain of the modified threonine has an outside orientation (for details, see chapter by I. Just and P. Boquet of this volume).

2.1.4 Glucosylation of GTPases Induces Dramatic Effects in Intoxicated Cells

The most prominent effect of the modification of the GTPases is the collapse of the ACSK. In fact, TcdB was the first bacterial protein toxin shown to induce such cytoskeletal breakdown (THELESTAM and BRÖNNEGÅRD 1980). The body of a cell intoxicated with TcdA or TcdB rounds up and protrusions radiating from the rounded cell remain attached to the substrate. Therefore, this morphological effect has been termed actinomorphic, arborizing, or neurite-like. Interestingly, strains 1470 and 8864 of the F serogroup of *C. difficile* (TcdA-negative) produce cytotoxins which induce a different type of cytopathic effect (CPE) (EICHEL-STREIBER et al. 1995). Because of a different substrate pattern specificity, this CPE rather resembles the rounding effect induced by the *C. sordellii* lethal toxin (TcsL) (Chaves-Olarte et al. 1999).

The collapse of the ACSK in cells exposed to TcdA and TcdB seems to arise from modification of Rho alone. Three lines of evidence support this statement: (a) the *C. botulinum* C3 exoenzyme which modifies only Rho (by ADP-ribosylation) induces a similar arborized CPE (SAITO and NARUMIYA 1997), (b) microinjection of glucosylated Rho mimics the CPE induced by the *C. difficile* toxins (JUST et al. 1995a), and (c) transient overexpression of Rho prevents the effect of TcdB (GIRY et al. 1995). The detailed molecular events controlling the ACSK collapse resulting from covalent modification of Rho are not known. This is largely due to incomplete knowledge of the normal regulatory functions of this molecule.

As mentioned above, Rho is not the only substrate of the *C. difficile* toxins. Cdc42 and Rac are modified to the same extent. Their inactivation should produce additional important changes in the intoxicated cell, ranging from interference with gene transcription (these GTPases control several extracellularly activated kinases) to blockade of the formation of dynamic actin-based structures (filopodia, membrane ruffles) (HALL 1998; CARON and HALL 1998).

2.2 TcdA and TcdB Differ with Respect to Cytotoxic Potencies

Its has been generally appreciated that TcdB is a much more efficient cytotoxin than TcdA. Intriguingly, TcdA is more potent in the induction of biological effects, at least in animal intestines (LYERLY and WILKINS 1995). This difference in biological activities is striking since both toxins are glucosyltransferases and modify almost identical substrates. The reason for the difference in cytotoxic activity was recently clarified when CHAVES-OLARTE and coworkers (1997) demonstrated that the enzyme activity of TcdA is very much lower than that of TcdB. Accordingly, the V_{max} of TcdA was shown to be five times lower than that of TcdB (CIESLA and BOBAK 1998). Furthermore, there is still a highly significant difference in cytotoxicity when the toxins are applied directly in the cytosol by microinjection (CHAVES-OLARTE et al. 1997). Besides this difference in enzymatic potency, the ability to bind to specific receptors on the cell surface also appears to contribute to the difference in cytotoxic potencies (CHAVES-OLARTE et al. 1997). The reasons for the differences in other biological effects are currently not known.

2.3 TcdA and TcdB Act on Most Cell Types

So far, the *C. difficile* toxins have been found to intoxicate practically every tested cell. However, the sensitivities of different cell types vary, probably because of variations in cell surface receptor densities. Furthermore, different cell types may respond with differing symptoms, depending on the roles played by the target GTPases in the particular cell type intoxicated. Some cellular effects which are especially relevant for the pathogenesis of *C. difficile* disease will be discussed in the next section. Table 1 lists examples of morphological features of diverse cell types treated with the toxins, as reported during the past decade (see FIORENTINI and THELESTAM 1991 for references to earlier studies). Several effects on cellular signaling pathways have been recently described by authors who used the toxins as cell biology tools (see THELESTAM et al. 1999 and chapter by I. Just and P. Boquet of this volume).

Table 1. Morphological effects of *C. difficile* toxins on various types of cells

Cells	Toxin	Cellular effect	Reference
T84	TcdA, TcdB	ACSK effects, barrier function	HECHT et al. 1988, 1992
CHO	TcdA	Ultrastructural changes	KUSHNARYOV and SEDMAK 1989
HEp-2	TcdA, TcdB	ACSK effects	FIORENTINI et al. 1989
HeLa, HEp-2	TcdA	ACSK effects, nuclear polarization	FIORENTINI et al. 1990
B cells	TcdB	Multinucleation	SHOSHAN et al. 1990
T cells (Jurkat)	TcdA	Multinucleation	FIORENTINI et al. 1992
Porcine endothelial	TcdA, TcdB	Cell rounding	MÜLLER et al. 1992
Human colonic pancreatic carcinoma	TcdA	Cell rounding	KUSHNARYOV et al. 1992
IEC-6	TcdA	Cell rounding, blebbing, ACSK effects	FIORENTINI et al. 1993
Don, T84	TcdA, TcdB	Actinomorphic CPE, rounding	CHAVES-OLARTE et al. 1997
IEC-6	TcdB	Apoptosis	FIORENTINI et al. 1998

ACSK, actin cytoskeleton; CPE, cytopathic effect.

The only cell types reported to resist TcdA and TcdB have mutations leading to deficiencies in either the endosomal acidification (MERION et al. 1983) or the biosynthesis of UDP-Glc (FLORES-DÍAZ et al. 1997). In such mutant cells the intoxication is prevented either because internalization of the toxins is prevented (FLORIN and THELESTAM 1986; HENRIQUES et al. 1987) or because the level of UDP-Glc is too low to support the glucosyltransferase reaction (CHAVES-OLARTE et al. 1996).

3 TcdA and TcdB in *C. difficile* Disease

Early studies indicated that TcdA causes fluid accumulation and a strong inflammation in animal intestinal loops while TcdB alone was without effect in such assays (earlier references in THELESTAM et al. 1997). Thus, TcdA was denoted as an enterotoxin and regarded as the major virulence factor, while TcdB, the more potent cytotoxin, was believed to be unimportant in the pathogenesis of *C. difficile* disease (earlier references in LYERLY and WILKINS 1995). From a teleological point of view, however, one would expect that TcdB plays a role (other than being a useful tool in cell biology) since the toxin is produced at a high cost of energy along with TcdA. As we shall see below, both toxins indeed appear important in the disease according to recent in vitro and in vivo studies.

3.1 Effects of TcdA and TcdB on Intestinal Epithelia and Isolated Cells

TcdB is also capable of eliciting enterotoxic symptoms in intestinal loop assays, provided a trace amount of TcdA is present. This suggested that the toxins may act synergistically in the clinical situation (LYERLY and WILKINS 1995). The following in vitro experiments support the notion that both *C. difficile* toxins can act directly on intestinal epithelial cells. Both toxins were shown to affect adversely the barrier function of polarized human intestinal epithelial T84 cells cultured and tested in vitro (HECHT et al. 1988, 1992). TcdA had a tenfold stronger effect than TcdB in this model, consistent with the more efficient binding of TcdA to T84 cells and also with its stronger cytotoxicity in these cells (CHAVES-OLARTE et al. 1997). Using unpolarized IEC-6 and polarized Caco-2 intestinal cells in an in vitro wound-induced migration assay, Santos and coworkers (1997) demonstrated that both toxins impaired cell migration in both cell lines. Thus, the toxin-induced inhibition of Rho function could be relevant in the disease mechanism, not only through the loss of the epithelial barrier function (NUSRAT et al. 1995), but also by preventing the cell migration which is essential for mucosal restitution (SANTOS et al. 1997).

Interestingly, Riegler and coworkers (1995) observed that in 5 hours both toxins caused mucosal necrosis and a decreased barrier function of mucosal strips isolated from normal human colonic epithelium. TcdB was ten times more potent than TcdA in causing these effects. Thus, the higher cytotoxic and enzymatic potency of TcdB was reflected in this very relevant in vitro model. More recently, the same authors demonstrated a reduction of these damaging effects by the epidermal growth factor (EGF) provided it was added to the serosal side of the colonic strips prior to treatment with toxin from the mucosal side (RIEGLER et al. 1997). Possibly, EGF stabilizes the ACSK by triggering signaling cascades that switch small GTPases into the GTP-loaded active state, thereby making them poorer substrates for the toxins.

3.2 Activation of the Immune System by TcdA and TcdB

A strong inflammatory reaction is typical in *C. difficile* colitis and several recent studies focus on direct and indirect effects of TcdA and TcdB on immune cells. Both toxins stimulated a release of tumor necrosis factor-α (TNF-α) from cultured monocytes, and TcdB was found to be approximately 1000 times more potent than TcdA in this system (SOUZA et al. 1997). The macrophage-derived TNF-α and lipoxygenase products in turn mediated an intense migration of neutrophils. Accordingly, both toxins were shown to activate monocytes in vitro to release interleukin-8 (IL-8), facilitating neutrophil extravasation and tissue infiltration (LINEVSKY et al. 1997). In keeping with these findings, the neutrophil migration evoked by TcdA in the peritoneal cavities of rats was shown to be partially dependent on macrophage-derived cytokines (ROCHA et al. 1997). Using mast cell-deficient mice, Pothoulakis and coworkers (1998) recently demonstrated the importance of mast cells for neutrophil recruitment and fluid secretion induced by TcdA in vivo. Furthermore, isolated mast cells were shown to respond to TcdA by releasing TNF-α (CALDERÓN et al. 1998). A prolonged incubation of the mast cells with TcdA also led to impairment of their functions and survival. Such effects might hamper the capacity of mast cells to counteract the infection, thus contributing to the prolonged pathogenic effects of the *C. difficile* toxins (CALDERÓN et al. 1998). Interestingly, IL-11 administered to rat ileal loops before TcdA reduced all the adverse intestinal effects observed in controls without IL-11 (CASTAGLIUOLO et al. 1997a). The authors speculated that IL-11 might inhibit the release of inflammatory mediators. In conclusion, several recent studies have shown that both *C. difficile* toxins can activate cells of the immune system.

3.3 Neuronal Activation by TcdA

Earlier experiments suggested that the effects of TcdA in animal intestines involve the neuronal system somehow (CASTAGLIUOLO et al. 1994; KELLY et al. 1994). More recently, the same authors observed that in rats suffering from TcdA-enteritis there are increased substance P (SP) responses both in dorsal root ganglia and in intestinal macrophages (CASTAGLIUOLO et al. 1997b). Injection of TcdA into rat ileum evoked an increased SP content in lumbar dorsal root ganglia and mucosal scrapings 30–60 min after toxin administration, i.e., well before the increased fluid secretion and mucosal necrosis occurred. Compared to control cells, the lamina propria macrophages obtained from TcdA-injected loops released greater amounts of TNF-α and SP, and pretreatment of the rats with an SP antagonist inhibited the TNF-α release (CASTAGLIUOLO et al. 1997b). In this context it is of interest to bear in mind the earlier observed massive increase of SP receptors in the small blood vessels and lymphoid aggregates in a bowel specimen of human pseudomembranous colitis (MANTYH et al. 1996). In conclusion, there is convincing evidence for a TcdA-induced activation of the enteric nervous system by an early transepithelial signal of unknown nature. Binding of TcdA to a receptor on the mucosal side of intestinal epithelial cells could induce the release of a cytokine(s) from the enterocytes which would diffuse into the lamina propria and activate primary sensory afferent neurons whose cell bodies are located in the dorsal root ganglia (POTHO-ULAKIS et al. 1998).

4 Model for Action of the *C. difficile* Toxins in the Intestine

Any reasonable model for the pathogenesis of *C. difficile* disease must accommodate the following observations:

1. TcdA and TcdB have one well-characterized enzymatic activity. The glucosyltransfer is the basis for their general cytotoxicity. TcdA is a far less potent glucosyltransferase and cytotoxin than TcdB.
2. TcdA alone is capable of eliciting intestinal fluid secretion and inflammation in animal models.
3. TcdB in itself has no enterotoxic effect in animal models. Subenterotoxic doses of TcdA synergize with TcdB to elicit enterotoxicity.

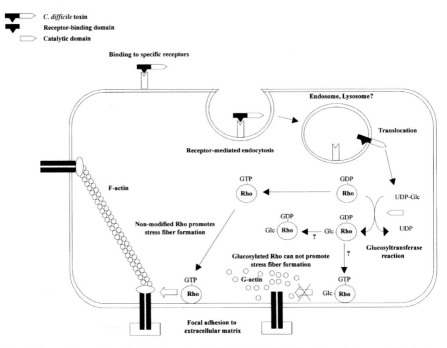

Fig. 1. General scheme for cellular intoxication by TcdA/TcdB. Note: the depicted cleavage of the toxins before release into the cytosol is hypothetical. Rho is shown here as the target for glucosylation, but Rac and Cdc42 are also targets for both toxins

4. Both toxins are capable of activating immune cells in vitro and in vivo to release various cytokines.
5. In the clinical situation, both toxins are produced in the lumen of the intestine. The target immune cells are located in the lamina propria on the serosal side. The receptors for TcdB appear to be more concentrated on the basolateral side of the intestinal epithelium.

Based on these considerations and the data reviewed above, the following scenario for the pathogenesis of *C. difficile* disease can be depicted (Fig. 2). The pathogenic process begins with the binding of TcdA to receptors on the mucosal side of intestinal cells. This binding elicits an early transepithelial signal of unknown nature. The signal induces the release of mediators on the basolateral side of the epithelium which in turn activate neurons that trigger fluid secretion. The same mediators also activate tissue macrophages to produce proinflammatory cytokines. These events lead to an opening of the tight junctions, a strong inflammatory cell infiltration in the epithelium, and later a mucosal injury. All this can take place in the absence of TcdB. In the clinical situation when TcdB is also present it moves across the epithelium via the opened tight junctions, reaching the basolateral cell surfaces. From here it can be more efficiently internalized and exert a direct cytotoxic effect on the enterocytes, strongly aggravating the mucosal necrosis and

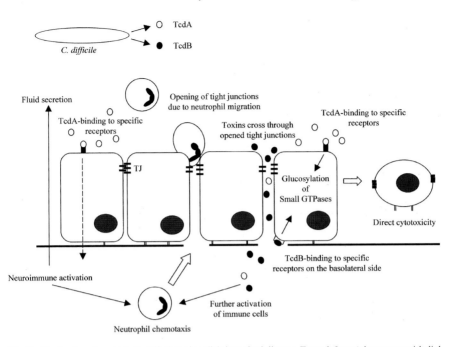

Fig. 2. Mechanisms by which *C. difficile* toxins elicit intestinal disease. From *left to right*: a transepithelial neuroimmune activation by TcdA leads to fluid secretion, mast cell activation, neutrophil migration, and opening of tight junctions. Both toxins can then move across the epithelium and exert their effects on immune cells. In addition, TcdB can also intoxicate enterocytes from the basolateral side, leading to cell rounding and detachment

inflammation. Here both toxins can also directly activate, and at high concentrations destroy, immune cells. All these events in concert will induce gross pathological changes, ultimately resulting in focal pseudomembrane formation.

The notion that TcdA, but not TcdB, is able to signal across the epithelium, thereby activating neurons and immune cells, gives a reasonable explanation for the strong effects in intestinal loop assays of TcdA alone, in spite of its low enzymatic potency. This signaling capacity might be connected with the larger C-terminal ligand domain of TcdA, which appears to be exposed on the surface of the native toxin molecule (SAUERBORN et al. 1994). In contrast, TcdB is likely to have a different 3D-configuration, also exposing N-terminal parts of the molecule, since its B-cell epitopes, in contrast to those of TcdA, appear distributed over the entire sequence of the molecule (SAUERBORN et al. 1994). The different types of biological effects observed with TcdA might be explained if it binds to different receptors, each mediating a different biological response as indirectly suggested by certain observations (e.g., SAUERBORN et al. 1997; FAUST et al. 1998; MAZUSKI et al. 1998). One receptor could be crucial for the transepithelial signaling, whereas TcdA binding to another receptor may trigger its uptake into enterocytes, enabling its action on GTPases.

Acknowledgements. Work performed in the authors' laboratory was supported by the Swedish Medical Research Council (05969), the Karolinska Institutet Research Funds, the Magnus Bergvall Foundation, and the Swedish Society for Medical Research. Esteban Chaves-Olarte is the recipient of a fellowship from the KIRT program supported by SIDA-Sarec.

References

Calderón GM, Torres-López J, Lin TJ, Chaves B, Hernández M, Munoz O, Befus AD, Enciso JA (1998) Effects of toxin A from *Clostridium difficile* on mast cell activation and survival. Infect Immun 66:2755–2761

Caron E, Hall A (1998) Identification of two distinct mechanisms of phagocytosis controlled by different Rho GTPases. Science 282:1717–1720

Castagliuolo I, LaMont JT, Letourneau R, Kelly CP, O'Keane JC, Jaffer A, Theoharides TC, Pothoulakis C (1994) Neuronal involvement in the intestinal effects of *Clostridium difficile* toxin A and *Vibrio cholerae* enterotoxin in rat ileum. Gastroenterol 107:657–665

Castagliuolo I, Kelly CP, Qiu BS, Nikulasson ST, LaMont JT, Pothoulakis C (1997a) IL-11 inhibits *Clostridium difficile* toxin A enterotoxicity in rat ileum. Am J Physiol 273:G333–343

Castagliuolo I, Keates AC, Qiu B, Kelly CP, Nikulasson S, Leeman SE, Pothoulakis C (1997b) Increased substance P responses in dorsal root ganglia and intestinal macrophages during *Clostridium difficile* toxin A enteritis in rats. Proc Natl Acad Sci USA 94:4788–4793

Chaves-Olarte E, Florin I, Boquet P, Popoff M, Eichel-Streiber Cv, Thelestam M (1996) UDP-Glucose deficiency in a mutant cell line protects against glucosyltransferase toxins from *Clostridium difficile* and *Clostridium sordellii*. J Biol Chem 271:6925–6932

Chaves-Olarte E, Weidmann M, Eichel-Streiber CV, Thelestam M (1997) Toxins A and B from *Clostridium difficile* differ with respect to enzymatic potencies, cellular substrate specificities, and surface binding to cultured cells. J Clin Invest 100:1734–1741

Chaves-Olarte E, Löw P, Freer E, Norlin T, Weidmann M, Eichel-Streiber CV, Thelestam M (1999) A novel cytotoxin from *Clostridium difficile* serogroup F is a functional hybrid between two other large clostridial cytotoxins. J Biol Chem 274:11046–11052

Ciesla WP, Bobak DA (1998) *Clostridium difficile* toxins A and B are cation-dependent UDP-glucose hydrolases with differing catalytic activities. J Biol Chem 273:16021–16026

Eichel-Streiber CV, Warfolomeow I, Knautz D, Sauerborn M, Hadding U (1991) Morphological changes in adherent cells induced by *Clostridium difficile* toxins. Biochem Soc Trans 19:1154–1160

Eichel-Streiber CV, Meyer ZU, Heringdorf M, Habermann E, Hartingen S (1995) Closing in on the toxic domain through analysis of a variant *Clostridium difficile* cytotoxin B. Mol Microbiol 17:313–321

Eichel-Streiber CV, Boquet P, Sauerborn M, Thelestam M (1996) Large clostridial cytotoxins – a family of glycosyltransferases modifying small GTP-binding proteins. Trends Microbiol 4:375–382

Faust C, Ye B, Song KP (1998) The enzymatic domain of *Clostridium difficile* toxin A is located within its N-terminal region. Biochem Biophys Res Commun 251:100–105

Fiorentini C, Thelestam M (1991) *Clostridium difficile* toxin A and its effects on cells. Toxicon 29:543–567

Fiorentini C, Arancia G, Paradisi S, Donelli G, Giuliano M, Piemonte F, Mastrantonio P (1989) Effects of *Clostridium difficile* toxins A and B on cytoskeleton organization in HEp-2 cells: a comparative morphological study. Toxicon 27:1209–1218

Fiorentini C, Malorni W, Paradisi S, Giuliano M, Mastrantonio P, Donelli G (1990) Interaction of *Clostridium difficile* toxin A with cultured cells: cytoskeletal changes and nuclear polarization. Infect Immun 58:2329–2336

Fiorentini C, Chow SC, Mastrantonio P, Jeddi-Tehrani M, Thelestam M (1992) *Clostridium difficile* toxin A induces multinucleation in the human leukemic T cell line JURKAT. Eur J Cell Biol 57: 292–297

Fiorentini C, Donelli G, Nicotera P, Thelestam M (1993) *Clostridium difficile* toxin A elicits Ca^{2+}-independent cytotoxic effects in cultured normal rat intestinal crypt cells. Infect Immun 61:3988–3993

Fiorentini C, Fabbri A, Falzano L, Fattorossi A, Matarrese P, Rivabene R, Donelli G (1998) *Clostridium difficile* toxin B induces apoptosis in intestinal cultured cells. Infect Immun 66:2660–2665

Flores-Diaz M, Alape-Girón A, Persson B, Pollesello P, Moos M, Eichel-Streiber CV, Thelestam M, Florin I (1997) Cellular UDP-glucose deficiency caused by a single point mutation in the UDP-glucose pyrophosphorylase gene. J Biol Chem 272:23784–23791

Florin I, Thelestam M (1986) Lysosomal involvement in cellular intoxication with *Clostridium difficile* toxin B. Microb Pathogen 1:373–385

Giry M, Popoff MR, Eichel-Streiber CV, Boquet P (1995) Transient expression of RhoA, -B, and -C GTPases in HeLa cells potentiates resistance to *Clostridium difficile* toxins A and B but not to *Clostridium sordellii* lethal toxin. Infect Immun 63:4063–4071

Hall A (1998) Rho GTPases and the actin cytoskeleton. Science 279:509–514

Hecht G, Pothoulakis C, LaMont JT, Madara J (1988) *Clostridium difficile* toxin A perturbs cytoskeletal structure and tight junction permeability of cultured human intestinal epithelial monolayers. J Clin Invest 82:1516–1524

Hecht G, Koutsouris A, Pothoulakis C, LaMont JT, Madara J (1992) *Clostridium difficile* toxin B disrupts the barrier function of T84 monolayers. Gastroenterol 102:416–423

Henriques B, Florin I, Thelestam M (1987) Cellular internalisation of *Clostridium difficile* toxin A. Microb Pathogen 2:455–463

Hofmann F, Busch C, Prepens U, Just I, Aktories K (1997) Localization of the glucosyltransferase activity of *Clostridium difficile* toxin B to the N-terminal part of the holotoxin. J Biol Chem 272:11074–11078

Just I, Selzer J, Wilm M, Eichel-Streiber CV, Mann M, Aktories K (1995a) Glucosylation of Rho proteins by *Clostridium difficile* toxin B. Nature 375:500–503

Just I, Wilm M, Selzer J, Rex G, Eichel-Streiber Cv, Mann M, Aktories K (1995b) The enterotoxin from *Clostridium difficile* (ToxA) monoglucosylates the Rho proteins. J Biol Chem 270:13932–13936

Kato H, Kato N, Watanabe K, Iwai N, Nakamura H, Yamamoto T, Suzuki K, Kim SM, Chong Y, Wasito EB (1998) Identification of toxin A-negative, toxin B-positive *Clostridium difficile* by PCR. J Clin Microbiol 36:2178–2182

Kelly CP, Pothoulakis C, LaMont JT (1994) *Clostridium difficile* colitis. N Engl J Med 330:257–262

Krivan HC, Clark GF, Smith DF, Wilkins TD (1986) Cell surface binding site for *Clostridium difficile* enterotoxin: evidence for a glycoconjugate containing the sequence Galα1-3Galβ1-4GlcNAc. Infect Immun 53:573–581

Kushnaryov VM, Sedmak JJ (1989) Effect of *Clostridium difficile* enterotoxin A on ultrastructure of Chinese hamster ovary cells. Infect Immun 57:3914–3921

Kushnaryov VM, Redlich PN, Sedmak JJ, Lyerly DM, Wilkins TD (1992) Cytotoxicity of *Clostridium difficile* toxin A for human colonic and pancreatic carcinoma cell lines. Cancer Res 52:5096–5099

Linevsky JK, Pothoulakis C, Keates S, Warny M, Keates AC, LaMont JT, Kelly CP (1997) IL-8 release and neutrophil activation by *Clostridium difficile* toxin-exposed human monocytes. Am J Physiol 273:G1333–1340

Lyerly DM, Wilkins TD (1995) *Clostridium difficile*. In: Blaser MJ et al. (eds) Infections of the Gastro-intestinal Tract. Raven Press Ltd., New York, pp. 867–891

Mantyh CR, Maggio JE, Mantyh PW, Vigna SR, Pappas TN (1996) Increased substance P receptor expression by blood vessels and lymphoid aggregates in *Clostridium difficile*-induced pseudomembranous colitis. Dig Dis & Sci 41:614–620

Mazuski JE, Panesar N, Tolman K, Longo WE (1998) In vitro effects of *Clostridium difficile* toxins on hepatocytes. J Surg Res 79:170–178

Merion M, Schlesinger P, Brooks RM, Moehring JM, Moehring TJ, Sly WS (1983) Defective acidification of endosomes in Chinese hamster ovary cell mutants cross-resistant to toxins and viruses. Proc Natl Acad Sci USA 80:5315–5319

Müller H, Eichel-Streiber CV, Habermann E (1992) Morphological changes of cultured endothelial cells after microinjection of toxins that act on the cytoskeleton. Infect Immun 60:3007–3010

Nusrat A, Giry M, Turner JR, Colgan SP, Parkos CA, Carnes D, Lemichez E, Boquet P, Madara JL (1995) Rho protein regulates tight junctions and perijunctional actin organization in polarized epithelia. Proc Natl Acad Sci USA 92:10629–1063

Pothoulakis C, Gilbert RJ, Cladaras C, Castaglivolo I, Semenza G, Hitti Y, Montcrief JS, Linevsky J, Kelly CP, Nikulasson S, Desai HP, Wilkins TD, LaMont JT (1996) Rabbit sucrase-isomaltase contains a functional intestinal receptor for *Clostridium difficile* toxin A. J Clin Invest 98:641–649

Pothoulakis C, Castagliuolo I, LaMont JT (1998) Nerves and intestinal mast cells modulate responses to enterotoxins. News Physiol Sci 13:58–63

Riegler M, Sedivy R, Pothoulakis C, Hamilton G, Zacherl J, Bischof G, Cosentini E, Feil W, Schiessel R, LaMont JT, Wenzl E (1995) *Clostridium difficile* toxin B is more potent than toxin A in damaging human colonic epithelium in vitro. J Clin Invest 95:2004–2011

Riegler M, Sedivy R, Sogukoglu T, Castagliuolo I, Pothoulakis C, Cosentini E, Bischof G, Hamilton G, Teleky B, Feil W, LaMont JT, Wenzl E (1997) Epidermal growth factor attenuates *Clostridium difficile* toxin A- and B-induced damage of human colonic mucosa. Am J Physiol 273:G1014–1022

Rocha MF, Maia ME, Bezerr LR, Lyerly DM, Guerrant RL, Ribeiro RA, Lima AA (1997) *Clostridium difficile* toxin A induces the release of neutrophil chemotactic factors from rat peritoneal macrophages: role of interleukin-1β, tumor necrosis factor alpha, and leukotrienes. Infect Immun 65:2740–2746

Rupnik M, Avesani V, Janc M, Eichel-Streiber CV, Delmée M (1998) A novel toxinotyping scheme and correlation of toxinotypes with serogroups of *Clostridium difficile* isolates. J Clin Microbiol 36:2240–2247

Saito Y, Narumiya S (1997) Preparation of *Clostridium botulinum* C3 exoenzyme and application of ADP-ribosylation of Rho proteins in biological systems. In: Aktories K (ed) Bacterial toxins – tools in cell biology and pharmacology. Chapman & Hall, Weinheim, pp 85–92

Santos MF, McCormack SA, Guo Z, Okolicany J, Zheng Y, Johnson LR, Tigyi G (1997) Rho proteins play a critical role in cell migration during the early phase of mucosal restitution. J Clin Invest 100:216–225

Sauerborn M, Hegenbarth S, Laufenberg-Feldmann R, Leukel P, Eichel-Streiber CV (1994) Monoclonal antibodies discriminating between *Clostridium difficile* toxins A and B. In: Freer J et al. (eds) Bacterial protein toxins. Gustav Fischer Verlag, Stuttgart, pp 510–511

Sauerborn M, Leukel P, Eichel-Streiber Cv (1997) The C-terminal ligand-binding domain of *Clostridium difficile* toxin A (TcdA) abrogates TcdA-specific binding to cells and prevents mouse lethality. FEMS Microbiol Lett 155:45–54

Shoshan MC, Åman P, Skoog S, Florin I, Thelestam M (1990) Microfilament-disrupting *Clostridium difficile* toxin B causes multinucleation of transformed cells but does not block capping of membrane Ig. Eur J Cell Biol 53:357–363

Souza MH, Melo-Filho AA, Rocha MF, Lyerly DM, Cunha FQ, Lima AA, Ribeiro RA (1997) The involvement of macrophage-derived tumour necrosis factor and lipoxygenase products on the neutrophil recruitment induced by *Clostridium difficile* toxin B. Immunology 91:281–288

Teneberg S, Lönnroth I, Torres Lopez JF, Galili U, Halvarsson MO, Ångström J, Karlsson KA (1996) Molecular mimicry in the recognition of glycosphingolipids by Galα3-Galβ4-GlcNAc-binding *Clostridium difficile* toxin A, a human natural anti α-galactosyl IgG and the monoclonal antibody Gal-13:characterization of a binding-active human glycosphingolipid, non-identical with the animal receptor. Glycobiology 6:599–609

Thelestam M, Brönnegård M (1980) Interaction of cytopathogenic toxin from *Clostridium difficile* with cells in tissue culture. Scand J Inf Dis Suppl 22:16–29

Thelestam M, Florin I, Chaves-Olarte E (1997) *Clostridium difficile* toxins. In: Aktories K (ed) Bacterial toxins – tools in cell biology and pharmacology. Chapman & Hall, Weinheim, pp 141–158

Thelestam M, Chaves-Olarte E, Moos M, Eichel-Streiber CV (1999) Clostridial toxins acting on the cytoskeleton. In: Alouf JE, Freer J (eds) Sourcebook of Bacterial Toxins. Academic Press (in press)

Tucker KD, Wilkins TD (1991) Toxin A of *Clostridium difficile* binds to the human carbohydrate antigens I, X and Y. Infect Immun 59:73–78

Large Clostridial Cytotoxins as Tools in Cell Biology

I. Just[1] and P. Boquet[2]

1 Introduction

Clostridium difficile toxin A and B, the lethal and haemorrhagic toxin from *Clostridium sordellii* and the α-toxin from *Clostridium novyi* are encompassed in the family of large clostridial cytotoxins (LCC) (VON EICHEL-STREIBER et al. 1996; BOQUET et al. 1998). This designation came from their molecular mass of about 300kDa and their obvious cytotoxic activity to induce disaggregation of the actin cytoskeleton. Despite their comparable toxic activities towards cultured cell lines, the toxins are produced by strains which are involved in distinct disease entities (HATHEWAY 1990). The best characterised are toxins A and B from *C. difficile*, which are of major clinical importance causing antibiotic-associated pseudomembranous colitis (KELLY et al. 1994; KELLY and LaMONT 1998).

2 The Family of Large Clostridial Cytotoxins

Large clostridial cytotoxins are intracellularly acting toxins which have to enter the intact cell through a sophisticated process. The toxins bind to specific membrane

---

[1] Institut für Pharmakologie und Toxikologie der Universität Freiburg, Hermann-Herder Strasse 5, 79104 Freiburg, Germany

[2] INSERM U 452, Faculté de Médecine, 28 Avenue de Valombrose, 06107 Nice, France

receptors which can be proteins, glycoproteins or glycolipids. Only for toxin A has a specific binding partner, the sucrase-isomaltase glycoprotein, been reported; the receptors of other LCCs are completely unknown (POTHOULAKIS et al. 1996). The binding to the membranous receptor induces endocytosis to deliver the toxins to acidic endosomal compartments. Decrease in pH is thought to induce refolding of the toxin which eventually results in translocation from the endosomes to the cytosolic compartment where the toxins exert their biological activity. These steps of cell entry and intracellular activity are reflected by the structure of the toxins. They are single-chain in structure with three distinct functional domains: (a) the C-terminally located receptor-binding domain, (b) the intermediary located putative translocation domain and (c) the N-terminal catalytic domain harbouring the mono-glucosyltransferase activity (LYERLY and WILKINS 1995; HOFMANN et al. 1997).

All members of the LCC family monoglucosylate Rho and Ras proteins, members of the superfamily of low molecular mass GTP-binding proteins. Rho/ Ras are upstream regulators in signal pathways governing cell differentiation, cell transformation and various aspects of the actin cytoskeleton (reviewed by ZOHN et al. 1998; VAN AELST and D'SOUZA-SCHOREY 1997). Whereas Rho and Ras are the preferred intracellular targets, other subfamilies such as Arf, Rab and Ran, which participate in intracellular trafficking, are not modified. The LCCs recruit the nucleotide sugar UDP-glucose (UDP-N-acetyl-glucosamine in the case of α-toxin) as cosubstrate and transfer the glucose (GlucNAc) moiety to a threonine receptor amino acid. The acceptor residue threonine is in position 37 of Rho and in the equivalent position (Thr-35) of Rac, Cdc42 and Ras, respectively (JUST et al. 1995a; POPOFF et al. 1996). Thr-37/35 resides in the so-called switch 1 which is essentially involved in the coupling of the active GTP-binding proteins to their respective effectors to transmit downstream signalling. Members of the LCC family show differences in their protein substrate specificity. The substrate specificity of toxin B, haemorrhagic toxin, toxin A and α-toxin is strictly confined to the Rho subfamily (Rho, Rac, Cdc42 and RhoG) whereas that of lethal toxin covers the Rho and Ras subfamilies (JUST et al. 1996; GENTH et al. 1996; POPOFF et al. 1996; SCHMIDT et al. 1998).

3 The Ras and Rho Subfamilies of GTP-Binding Proteins and Their Cellular Functions

The Ras superfamily of small GTP-binding proteins implicated in intracellular signal transduction are defined by their size (20–25kDa) and the fact that they share a minimum of 30% amino acid identity (CHARDIN 1993). This superfamily of proteins is divided into five main branches (Ras, Rho, Rab, Ran and ARF) according to their homology and their implication in the regulation of specific cellular processes.

Ras is the prototype of a regulatory protein involved in intracellular signalling. A main characteristic of Ras, shared by other small GTP-binding proteins, is the absolute requirement of membrane attachment to exert its activity (DOWNWARD 1990). Due to covalent modification by a farnesyl and a palmitate moiety to its carboxy-terminal domain (named the CAAX box), Ras is located to the cyto-plasmic face of the cell membrane (HANCKOCK et al. 1989; DOWNWARD 1990). The crystal structure of Ras (PAI et al. 1990) shows the structural domains involved in nucleotide binding and in GTP hydrolysis. Two domains of Ras undergo confor-mational change from the GDP-bound to the GTP-bound state. These domains have been called "switches" (WITTINGHOFER and VALENCIA 1995). Switch 1 (resi-dues 32–40) is involved in the interaction with the downstream effector Raf kinase (NASSAR et al. 1995). Switch 2 (residues 60–76) is implicated in the hydrolysis of GTP. Membrane localisation, binding of GTP, guanine nucleotide-dependent conformational changes, and hydrolysis of GTP are the key features of the Ras superfamily of GTP-binding proteins.

The cellular function of Ras is characterised by its activation/deactivation cycle. Through its ability to be turned on for a certain period of time, Ras is even an intracellular timer. Ras, in this task, uses the gamma phosphate of GTP to induce structural changes in both switch 1 and 2 domains. In addition to this mechanism, two groups of regulatory proteins which interact with Ras govern the activation/deactivation cycle. A factor named GEF (guanine nucleotide exchange factor), re-moves the bound GDP from inactivated Ras. Since there is a higher concentration of GTP than GDP in the cytosol, there is a rapid binding of GTP to Ras. To turn Ras · GTP into the inactive state, the GTP is hydrolysed. However, the intrinsic GTPase activity is very poor and must be accelerated by proteins named GAPs (GTPase activating proteins). The binding of a water molecule to Ras residue glu-tamine-61 (located in the switch 2 domain) is pivotal for GTP hydrolysis. However, this residue in Ras is not correctly positioned in the absence of GAP to achieve the transition state necessary for fast hydrolysis of GTP into GDP (SCHEFFZEK et al. 1997). Ras · GAP, by introducing an arginine residue [the so-called arginine finger (BOURNE 1997)] into the Ras molecule, correctly relocalises the glutamine residue close to the gamma phosphate of GTP thereby accelerating the hydrolysis.

The main Ras downstream effector is Raf. Raf, a serine/threonine kinase, is activated by Ras · GTP by binding to the membrane, phosphorylating the double specificity threonine/tyrosine kinases MEK1 and MEK2. The MEKs in turn phosphorylate ERK 1 and 2, which are broad specificity serine/threonine kinases. Upon phosphorylation, ERKs can translocate in the nucleus thereby activating either proliferation or differentiation, depending on the time ERKs are phospho-rylated. It is now clear that Ras, in addition to Raf, has other downstream effectors. Several lines of evidence indicate that Rho GTPases are Ras downstream targets at least for proliferation (SOHN et al. 1998). One mechanism for the co-operation between the Ras and the Rho family may involve the activation of Rac through phosphatidylinositol 3-phosphate (PIP_3) production, mediated by activation of the p110 catalytic subunit of the PI-3 kinase by Ras·GTP (RODRIGEZ-VICIANA 1994, 1995).

The Rho family of small GTP-binding protein encompasses 15 members [Rho A, B, C, D, E/(Rnd3), G, Rnd1(Rho6), Rnd2(Rho7), Rac1, Rac2, Rac3, Cdc42(brain), Cdc42(placenta), TC10 and TFF]. The effects on the cytoskeleton by Rho, Rac and Cdc42 have been fully documented on Swiss-3T3 cells. Rho activation leads to assembly of stress fibres and focal contacts (RIDLEY and HALL 1992). Cdc42 bound to GTP results in the development of filopodia whereas Rac activation produces membrane ruffling (RIDLEY et al. 1992; NOBES et al. 1995; KOZMA et al. 1995). In Swiss-3T3 fibroblasts, Rho GTPases have been shown to be sequentially activated. Cdc42 leads to activation of Rac which in turn stimulates Rho (NOBES and HALL 1995). However, this crosstalk between Cdc42, Rac and Rho has recently been challenged by several studies (SANDERS et al. 1999; SELLS et al. 1999; SANDER et al. 1999). Indeed, whereas Rho activation does not affect Rac or Cdc42 activities, Cdc42 and Rac downregulate Rho. Activation of Rac in NIH-3T3 fibroblasts blocks the ability of these cells to move (SANDER et al. 1999). Furthermore, in NIH-3T3 fibroblasts, activation of Rac allows establishment of cell–cell contacts (SANDER et al. 1999). The same activity of Rac on Rho was also shown for Cdc42 (SANDER et al. 1999). Thus a balance between Rho GTPases determines both the cellular morphology and the migratory activity (SANDER et al. 1999). LCCs, particularly toxin B, together inactivate Cdc42, Rac and Rho. These toxins will therefore inhibit a complicated array of cross regulations between Rho GTPases.

4 The Effects of LCCs on the Function of Rho and Ras GTPases

The functional effects of glucosylation of effector domain residue Thr-37/35 was most intensively studied with Ras glucosylated by lethal toxin (HERRMANN et al. 1998). Whereas nucleotide binding is not significantly altered, guanine-nucleotide exchange factor (Cdc25)-catalysed GTP loading is decreased but not completely inhibited. Modified Ras does not sequester the exchange factor like dominant negative mutants do. Intrinsic GTPase activity is markedly decreased and GTPase stimulation by GTPase activating protein p120^{GAP} and neurofibromin NF-1 is completely blocked, caused by failure of binding to glucosylated Ras. This alteration of the Ras GTPase cycle leads to an entrapment of glucosylated Ras in the GTP-bound state. However, the crucial step in downstream signalling, the Ras-effector coupling, is completely blocked. In conclusion, monoglucosylation has diverse effects on the Ras GTPase cycle, whereby the dominant action is the inhibition of the Ras-Raf coupling leading to complete blockade of Ras downstream signalling. The effects of glucosylation on the GTPase cycle were also found for Rho (SEHR et al. 1998), therefore it seems that the functional changes of the GTP-binding protein are general effects. The activation–deactivation cycle of Rho which is driven by the GTPase cycle is accompanied by a cycling between cytosol and membranous compartments, a unique feature for the Rho subfamily (ZOHN et al. 1998; SASAKI and TAKAI 1998; BOKOCH and DER 1993). The inactive state of Rho is characterised by complexation to the guanine-nucleotide dissociation inhibitor (GDI), which prevents loading with GTP and keeps Rho in the inactive form. Active GTP-bound Rho is translocated to the membranes to interact with effector proteins. After inactivation by GAP, the Rho · GDP is extracted by GDI and the complex enters the cytosolic pool of inactive Rho. Glucosylation dramatically alters this spatial cycling (GENTH et al. 1999). Glucosylated Rho is bound to the Rho-binding sites at membranes but is incapable of downstream signalling. Because of failure to bind to GDI, glucosylated Rho is entrapped at the membrane binding sites thereby inhibiting membrane translocation of unmodified cellular Rho. Thus, glucosylation has redundant effects on the GTPase cycle and on the cytosol–membrane cycling.

That the cytotoxic property of LCCs is mediated through their inherent glucosylating activity has been nicely shown by two experiments. Microinjection of the catalytic domain (covering amino acids 1–546), but not of one of the other domains, induces the same morphological and cytoskeletal features as the holotoxin does when added extracellularly (HOFMANN et al. 1997). Furthermore, microinjection of glucosylated Rho also causes actin cytoskeleton breakdown (JUST et al. 1995a).

In addition to the highly conserved effector binding domain (switch 1), accessory effector binding domains located C-terminally to switch 1 have been identified (ZONG et al. 1999; FUJISAWA et al. 1998). These additional interacting domains define the effector specificity, i.e. they determine why Rho couples to a different effector than Rac or Cdc42 thereby activating different signal pathways. However, switch 1 seems to be indispensable for effector coupling. Thus, glucosylation of switch 1 very likely blocks the complete array of downstream signalling of the modified GTPase. During the course of intoxication (glucosylation) it is conceivable that only some signal pathways are blocked, based on findings that the activation of some pathways need more Rho recruitment than others. The LCCs are designated as cytotoxins because morphological changes are easy to detect. But based on the current knowledge they cannot be classified as simple cytoskeleton disrupting toxins any longer; they indeed interfere with many vital functions of the cell, especially those assigned to the Rho and Ras proteins.

5 Susceptibility of Different Cell Lines to LCCs

All cell lines studied so far are sensitive to the LCCs although there are great differences in sensitivity. For example, Chinese hamster ovary (CHO) cells are highly sensitive to toxin B acting at a femto-molar range, whereas HEp-2 cells are quite insensitive; most fibroblasts and epithelial cells are insensitive to lethal toxin, but RBL (mast) cells are highly sensitive. Thus, the LCCs have access to cultured cell lines, primary cell culture or isolated cells from organs. For biochemical studies based on cell lines, it is essential to affect more than 90% of the cells to get measurable effects. This necessity is met by the LCCs but not by transfection

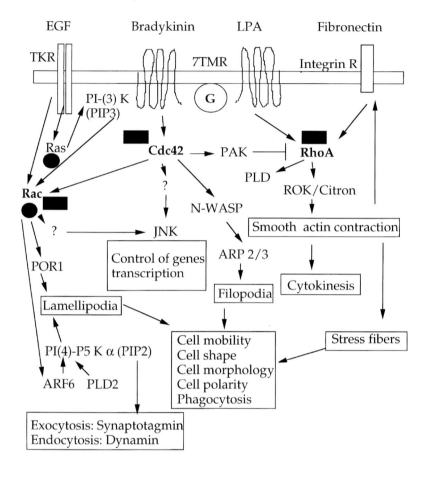

● *Clostridium sordellii* lethal toxin (LT)

■ *Clostridium difficile* toxin B (TCdB)

Fig. 1. How large clostridial toxins interfere and can be used to examine fundamental cell biology processes such as, cell mobility, cell shape, cell polarity, cell morphology, exocytosis, endocytosis, cytokinesis, phagocytosis and gene transcription through inhibition of Rho, Rac and Cdc42 GTPases. Rac is controlled by tyrosine kinase receptors (*TKR*) which are stimulated by growth factors, such as epithelial growth factor (*EGF*). Cdc42 is controlled by seven transmembrane receptors (*7TM*) linked to hetero-trimeric GTP-binding proteins (*G*). RhoA is controlled by *7TM* that binds the growth factor lysophos-phatidic acid (*LPA*) and integrin receptors (*Int R*) which recognise proteins of the extracellular matrix, such as fibronectin. Activation of Rac by Ras through stimulation of the phosphatidylinositol 3-phos-phate kinase [*PI(3)-K*] produces the lipid phosphatidylinositol 3-phosphate (*PIP₃*). PIP₃ activates the Rac nucleotide exchange factor, probably by binding to the plekstrin homology domain 3 (*PH3*) of Rac exchange factor. This is one (among four other possibilities) of the mechanisms which might link Ras to Rho GTPases (reviewed by ZOHN et al. 1998). Rac and Cdc42 activate transcription in the nucleus through stimulation of the Jun kinase pathway (VOJTEK and COOPER 1995). Rac is involved in membrane ruffling (lamellipodia) through simultaneous activation of POR1 (D'SOUZA-SCHOREY et al. 1997) and stimulation of ARF6 exchange factor (FRANCO et al. 1999). ARF6GTP, in conjunction with phosphatidic

techniques. Furthermore, the toxins are easy to apply and do not need incubation times longer than 24h. The cellular functions affected after toxin application can easily be assigned to the protein substrates of the toxin used.

However, one has to be careful with oversimplified interpretation of the results. The substrate specificities of the LCCs seem to be clear but indeed only RhoA has been shown to be an in vivo substrate. All the other protein substrates have been identified by testing glucosylation of known recombinant low molecular mass GTPases. The diverse substrate specificity of the isoforms of lethal toxin clearly shows that checking only one single prototype of a GTPase subfamily does not exclude the other members of the subfamily as substrate (POPOFF et al. 1996; HOFMANN et al. 1996; JUST et al. 1996). Furthermore, putative cellular targets may escape the identification through [^{14}C]glucosylation because they are expressed at a low level compared to the highly expressed Rho/Rac/Cdc42 proteins.

6 The Large Clostridial Cytotoxins as Tools to Study Cellular Functions

The LCCs are applied as tools to study cellular functions, i.e. their alteration. Because inactivation of Rho functions through glucosylation results in disaggregation of the actin cytoskeleton, it has to be checked whether the studied function is merely regulated by the actin filament system. To this end, toxins directly acting on actin thereby circumventing signal cascades are applicable. Two such toxins are available, cytochalasin D and the binary C2 toxin from *Clostridium botulinum*. Cytochalasin D easily permeates plasma membranes and induces depolymerisation of F-actin by direct binding to actin. The working concentration is 1–5μM (COOPER 1987). The C2 toxin consists of the enzyme component C2I and the binding component C2II which is only active after trypsin treatment. C2II translocates C2I through receptor-mediated endocytosis into the cytoplasm where C2I mono-ADP-

---

acid produced by phospholipase D2, activates the isoform alpha 2 of phosphatidylinositol 4–5 kinase [*PI-(4) 5K*] (HONDA et al. 1999). Stimulation of *PI-(4) 5K* leads to the production of the lipid phosphatidylinositol 4,5-biphosphate (*PIP$_2$*) allowing, probably by activation of gelsolin, membrane ruffling. The production of PIP$_2$ by ARF6 and Rac probably also controls regulated exocytosis through modulation of molecules, such as the calcium binding protein synaptotagmin (SCHIAVO et al. 1996), and endocytosis via molecules, such as dynamin, harbouring plekstrin homology 2 domains (*PH2*) involved in clathrin-mediated endocytosis (ACHIRILOAIE et al. 1999). Cdc42GTP activates the neuronal-Wiskott Aldrich syndrome protein (*N-WASP*) (MIKI et al. 1998). Activated N-WASP binds the actin related protein 2/3 (*ARP 2/3*) which is a protein complex allowing nucleation and elongation of actin filaments forming filopodia (ROHATGI et al. 1999). Cdc42 also stimulates the p21-activated kinase (*PAK*) which (probably by phosphorylating the light myosin chain) inhibits RhoA effects on the p160 Rho kinase (*ROK*) (SANDERS et al. 1999). Citron is related to ROK and implicated in midbody formation during cytokinesis (MADAULE et al. 1998). The cellular targets of *Clostridium sordellii* lethal (*closed circle*) and *Clostridium difficile* B (*closed rectangle*) toxins are indicated to demonstrate how these molecules can be used for cell biology studies

ribosylates G-actin. ADP-ribosylated actin is incapable of polymerisation and eventually leads to complete depolymerisation of the actin filaments (AKTORIES and JUST 1990; AKTORIES et al. 1997). The working concentration is about 200ng/ml for each component and the incubation time depends on the cell type but is in the range of 2–5h. If cytochalasin D or C2 toxin have the same effect as the LCCs, it is likely that the cellular function studied is affected or even regulated by the actin cytoskeleton.

The LCC-induced morphological changes are an excellent read-out to check whether the toxin is active on the cell type studied. However, in non-adherent cells, changes in cell-shape are not detectable. The visualisation of the sub-membranous actin cytoskeleton is applicable but gives no strong evidence for intracellular toxin activity. The only reliable method is to perform the differential glucosylation. If the toxin glucosylates the Rho proteins in the intact cell, a second [^{14}C]glucosylation of the lysate results in a decreased or even abolished incorporation of glucose. A less expensive technique is the usage of the C3-catalysed [^{32}P]ADP-ribosylation which, however, only proves modification of Rho but not of Rac or Cdc42. The differential glucosylation is also applicable to estimate the amount of inactivated (glucosylated) Rho proteins.

The LCC family encompasses five main members and, in addition, various isoforms. Every toxin from this family can be principally used as tool in cell biology but two members are advantageous for application. The most active toxin of the family is toxin B, whose use saves a lot of material and allows a reduced intoxication (incubation) time. Toxin B covers the Rho proteins (Rho, Rac, Cdc42, RhoG) as cellular targets. The lethal toxin is less active on most cell lines but its application has the advantage of distinguishing between Rac and Rho/Cdc42 because only Rac is modified. Furthermore, lethal toxin modifies also the Ras subfamily. Thus, by two experiments, it can be concluded whether Rho or Ras proteins, or whether Rho/Cdc42 or Rac are involved in the cellular functions studied.

In order to summarise the possible utilisation of toxin B or lethal toxin as tools in cell biology studies we have grouped, in Fig. 1, cell regulation pathways affecting cell polarity, cell morphology, cell shape, cytokinesis, control of transcription, endocytosis and exocytosis in which Rho, Rac and Cdc42 are pivotal or important players.

7 Purification and Conservation of LCCs

Toxin B and the lethal toxin are produced from the culture supernatant of the clostridia by ammonium sulphate precipitation followed by ion exchange chromatography (for details see JUST et al. 1997). For long-term storage the LCCs should be frozen at −80°C in the presence of 20% of glycerol. Repeated thawing and freezing should be strictly avoided because of dramatic loss of activity. For short-term storage, the toxins are stable at 4°C for 2–3 weeks.

Safety guidelines: The toxins are absorbed by mucous membranes, therefore, any contact by ingestion or inhalation is strictly to be avoided. The toxin waste is inactivated by autoclaving.

References

Achiriloaie M, Barylko B, Albanesi JP (1999) Essential role of the dynamin pleckstrin homology domain in receptor-mediated endocytosis. Mol Cell Biol 19:1410–1415

Aktories K, Just I (1990) Botulinum C2 toxin. In: Moss J, Vaughan M (eds) ADP-ribosylating toxins and G-proteins. American Society for Microbiology, Washington, D.C., pp 79–95

Aktories K, Prepens U, Sehr P, Just I (1997) Probing the actin cytoskeleton by Clostridium botulinum C2 toxin and Clostridium perfringens iota toxin. In: Aktories K (ed) Bacterial toxins. Chapman & Hall, Weinheim, pp 129–139

Bokoch GM, Der CJ (1993) Emerging concepts in the Ras superfamily of GTP-binding proteins. FASEB J 7:750–759

Boquet P, Munro P, Fiorentini C, Just I (1998) Toxins from anaerobic bacteria: specificity and molecular mechanisms of action. Current Opinion in Microbiology 1:66–74

Bourne HR (1997) The arginine finger strikes again. Nature 389:673–674

Chardin P (1993) In: Dicken BF, Birnbaumer L (eds) GTPases in biology. Springer-Verlag, Heidelberg, pp 159–176

Cooper JA (1987) Effects of cytochalasin and phalloidin on actin. J Cell Biol 105:1473–1478

Downward J (1990) The Ras superfamily of small GTP-binding proteins. Trends Biochem Sci 15: 449–477

D'Souza-Schorey C, Boshans RL, McDonough M, Stahl PD, Van Aelst L (1997) A role for POR1, a Rac1-interacting protein, in ARF6-mediated cytoskeletal rearrangements. EMBO J 16:5445–5454

Franco M, Peters PJ, Boretto J, van Donselaar E, Neri A, D'Souza-Schorey C, Chavrier P (1999) EFA6, a sec7 domain-containing exchange factor for ARF6, coordinates membrane recycling and actin cytoskeleton organization. EMBO J 18:1480–1491

Fujisawa K, Madaule P, Ishizaki T, Watanabe G, Bito H, Saito Y, Hall A, Narumiya S (1998) Different regions of Rho determine Rho-selective binding of different classes of Rho target molecules. J Biol Chem 273:18943–18949

Genth H, Aktories K, Just I (1999) Monoglucosylation of RhoA at Threonine-37 blocks cytosol-membrane cycling. J Biol Chem 274:29050–29056

Genth H, Hofmann F, Selzer J, Rex G, Aktories K, Just I (1996) Difference in protein substrate specificity between hemorrhagic toxin and lethal toxin from Clostridium sordellii. Biochem Biophys Res Commun 229:370–374

Hanckock JC, Paterson HF, Marshall CJ (1989) All ras proteins are polyisoprenylated but only some are palmitoylated. Cell 57:1167–1177

Honda A, Nogami M, Yokozeki T, Yamazaki M, Nakamura H, Watanabe H, Kawamoto K, Nakayama K, Morris AJ, Frohman MA, Kanaho Y (1999) Phosphatidylinositol 4-phosphate 5-kinase is a downstream effector of the small G protein ARF6 in membrane ruffle formation. Cell 99:521–532

Hatheway CL (1990) Toxigenic clostridia. Clin Microbiol Rev 3:66–98

Herrmann C, Ahmadian MR, Hofmann F, Just I (1998) Functional consequences of monoglucosylation of H-Ras at effector domain amino acid threonine-35. J Biol Chem 273:16134–16139

Hofmann F, Busch C, Prepens U, Just I, Aktories K (1997) Localization of the glucosyltransferase activity of Clostridium difficile toxin B to the N-terminal part of the holotoxin. J Biol Chem 272:11074–11078

Hofmann F, Rex G, Aktories K, Just I (1996) The Ras-related protein Ral is monoglucosylated by Clostridium sordellii lethal toxin. Biochem Biophys Res Commun 227:77–81

Just I, Selzer J, Hofmann F, Aktories K (1997) Clostridium difficile toxin B as a probe for Rho GTPases. In: Aktories K (ed) Bacterial toxins – tools in cell biology and pharmacology. Chapman & Hall, Weinheim, pp 159–168

Just I, Selzer J, Hofmann F, Green GA, Aktories K (1996) Inactivation of Ras by Clostridium sordellii lethal toxin-catalyzed glucosylation. J Biol Chem 271:10149–10153

Just I, Selzer J, Wilm M, Von Eichel-Streiber C, Mann M, Aktories K (1995a) Glucosylation of Rho proteins by *Clostridium difficile* toxin B. Nature 375:500–503

Just I, Wilm M, Selzer J, Rex G, Von Eichel-Streiber C, Mann M, Aktories K (1995b) The enterotoxin from *Clostridium difficile* (ToxA) monoglucosylates the Rho proteins. J Biol Chem 270:13932–13936

Kelly CP, LaMont JT (1998) *Clostridium difficile* infection. Annu Rev Med 49:375–390

Kelly CP, Pothoulakis C, LaMont JT (1994) *Clostridium difficile* colitis. New England J Med 330(4): 257–262

Kozma R, Ahmed S, Best A, Lim L (1996) The GTPase activating protein n-chimaerin cooperates with Rac1 and Cdc42Hs to induce the formation of lamellipodia and filopodia. Mol Cell Biol 16:5069–5080

Lyerly DM, Wilkins TD (1995) *Clostridium difficile*. In: Blaser MJ, Smith PD, Ravdin JI, Greenberg HB, Guerrant RL (eds) Infections of the Gastrointestinal Tract. Raven Press Ltd., New York, pp 867–891

Madaule P, Eda M, Watanabe N, Fujisawa K, Matsuoka T, Bito H, Ishizaki T, Narumiya S (1998) Role of citron kinase as a target of the small GTPase Rho in cytokinesis. Nature 394:491–494

Miki H, Sasaki T, Takai Y, Takanawa T (1998) Induction of filopodium formation by a WASP-related actin depolymerizing protein N-WASP. Nature 391:93–96

Nassar N, Horn G, Heurman C, Scheter A, Mc Cormick F, Wittinghofer A (1995) The 2.2 A crystal structure of the Ras-binding domain of the serine/threonine kinase C Raf in complex with Rap1A and a GTP analogue. Nature 375:554–560

Nobes CD, Hall A (1995) Rho, Rac, Cdc42 GTPases regulate the assembly of multimolecular focal complexes associated with actin stress fibres lamellipodia and filopodia. Cell 81:1–20

Popoff MR, Chaves OE, Lemichez E, Von Eichel-Streiber C, Thelestam M, Chardin P, Cussac D, Chavrier P, Flatau G, Giry M, Gunzburg J, Boquet P (1996) Ras, Rap, and Rac small GTP-binding proteins are targets for *Clostridium sordellii* lethal toxin glucosylation. J Biol Chem 271:10217–10224

Pai EF, Krengel U, Petsko GA, Goody RS, Kabsch W, Wittinghofer A (1990) Refined crystal structure of the triphosphate conformation of the H-ras p21 at 1,35A resolution: implication for the mechanism of GTP hydrolysis. EMBO J 9:2351–2359

Pothoulakis C, Gilbert RJ, Cladaras C, Castagliuolo I, Semenza G, Hitti Y, Montcrief JS, Linevsky J, Kelly CP, Nikulasson S, Desai HP, Wilkins TD, LaMont JT (1996) Rabbit sucrase-isomaltase contains a functional intestinal receptor for *Clostridium difficile* toxin A. J Clin Invest 98:641–649

Ridley A, Hall A (1992) The small GTP-binding protein Rho regulates the assembly of focal adhesion and actin stress fibres in response to growth factor. Cell 70:389–399

Ridley A, Paterson HF, Johnston CL, Dieckman O, Hall A (1992) The small GTP-binding protein Rac regulates growth factor-induced membrane ruffling. Cell 70:401–410

Rohatgi R, Ma L, Miki H, Lopez M, Kirchhausen T, Takenawa T, Kirschner MW (1999) The interaction between N-WASP and the Arp2/3 complex links Cdc42-dependent signals to actin assembly. Cell 97:221–231

Sander EE, ten Klooster JP, Van Delft S, Van der Kammen A, Collard JG (1999) Rac downregulates Rho activity: reciprocal balance between both GTPases determines cellular morphology and migratory behaviour. J Cell Biol 147:1009–1021

Sanders LC, Matsumura F, Bokoch GM, de Lanerolle P (1999) Inhibition of myosin light chain kinase by p21-activated kinase. Science 283:2083–2085

Sasaki T, Takai Y (1998) The Rho small G protein family-Rho GDI system as a temporal and spatial determinant for cytoskeletal control. Biochem Biophys Res Commun 245:641–664

Sells MA, Boyd JT, Chernoff J (1999) P21-activated kinase 1 (PAK1) regulates cell mobility in mammalian fibroblasts. J Cell Biol 145:837–849

Scheffzek K, Ahmadian MR, Kabsch W, Wiesmüller L, Lautwein A, Schmitz F, Wittinghofer A (1997) The Ras GAP complex: structural basis for GTPase activation and its loss in oncogenic Ras mutant. Science 277:333–338

Schiavo G, Qu-Ming Gu, Prestwich GD, Söllner TH, Rothman JE (1996) Calcium-dependent switching of the specificity of phosphoinositide binding to synaptotagmin. PNAS 93:13327–13332

Schmidt M, Vo M, Thiel M, Bauer B, Grannass A, Tapp E, Cool RH, De Gunzburg J, Von Eichel-Streiber C, Jakobs KH (1998) Specific inhibition of phorbol ester-stimulated phospholipase D by *Clostridium sordellii* lethal toxin and *Clostridium difficile* toxin B-1470 in HEK-293 cells. J Biol Chem 273:7413–7422

Sehr P, Joseph G, Genth H, Just I, Pick E, Aktories K (1998) Glucosylation and ADP-ribosylation of Rho proteins – effects on nucleotide binding, GTPase activity, and effector-coupling. Biochemistry 37:5296–5304

Van Aelst L, D'Souza-Schorey C (1997) Rho GTPases and signaling networks. Genes Dev 11:2295–2322

Vojtek AB, Cooper JA (1995) Rho family members: activators of MAP kinase cascades. Cell 82:527–529

Von Eichel-Streiber C, Boquet P, Sauerborn M, Thelestam M (1996) Large clostridial cytotoxins – a family of glycosyltransferases modifying small GTP-binding proteins. Trends Microbiol V4:375–382

Wittinghofer A, Valencia A (1995) Three dimensional structure of Ras and the ras-related proteins. In: Zerial M, Huber EH (eds) Guidebook to the small GTPases. Oxford University Press, New York, pp 20–29

Zohn IM, Campbell SL, Khosravi-Far R, Rossman KL, Der CJ (1998) Rho family proteins and Ras transformation: the RHOad less traveled gets congested. Oncogene 17:1415–1438

Zong H, Raman N, Mickelson-Young LA, Atkinson SJ, Quilliam LA (1999) Loop 6 of RhoA confers specificity for effector binding, stress fiber formation, and cellular transformation. J Biol Chem 274:4551–4560

Pathogenesis and Clinical Manifestations
of *Clostridium difficile* Diarrhea and Colitis

R.J. Farrell and J.T. LaMont

1 Introduction

Clostridium difficile is a spore-forming, gram-positive anaerobic bacillus discovered in 1935 as a commensal organism in the fecal flora of healthy newborn infants (Hall and O'Toole 1935). The organism was given its name because it grew very slowly in culture and was difficult to isolate in pure culture. Its presence in the stool of 50% of healthy neonates suggested that *C. difficile* was a commensal not a pathogen, even though it was toxigenic in broth culture. Not long after its original description, *C. difficile* passed into obscurity until its rediscovery some four decades later. Antibiotic-associated pseudomembranous colitis became prevalent in the 1960s and 1970s with the introduction into clinical practice of broad-spectrum antibiotics. The frequent association of clindamycin and lincomycin therapy with pseudomembranous colitis led to the term "clindamycin colitis" (Tedesco et al.

Division of Gastroenterology, Beth Israel Deaconess Medical Center, Harvard Medical School, 330 Brookline Avenue, Boston, MA 02215, USA

1974). A breakthrough occurred in 1978 when *C. difficile* was identified as the source of a cytotoxin found in the stool of patients with pseudomembranous colitis (BARTLETT et al. 1978). During the two decades since its rediscovery, a great deal has been learned about the pathophysiology, epidemiology, and management of *C. difficile* diarrhea, yet many challenges remain. At present this organism continues to infect up to 20% of individuals admitted to hospital, making *C. difficile* colitis the most common enteric infection in hospital practice (KELLY et al. 1994).

2 Pathogenesis

The normal human colonic microflora of adults and of children over 2 years old is capable of preventing colonization by *C. difficile*. As shown in Fig. 1, the pathogenesis of *C. difficile* diarrhea involves an initial disruption of the normal colonic bacterial flora by antibiotics, allowing colonization with *C. difficile* to occur if the individual is exposed to the organism or its spores. The specific organism or group of organisms of normal adult microflora which exclude *C. difficile* is not entirely clear, but anaerobic species, including *Bacteroides*, may be especially important. For example, treatment with lyophilized *Bacteroides* species can inhibit the growth of *C. difficile* in the stools of patients with chronic recurrent infection (TVEDE and RASK-MADSEN 1989). This infection is typically acquired in hospitals because approximately 30–50% of patients receive antibiotics in an environment where *C. difficile* is highly prevalent. As a result, *C. difficile* infection causes colitis and diarrhea in approximately one in ten hospitalized patients in the United States (MCFARLAND et al. 1989).

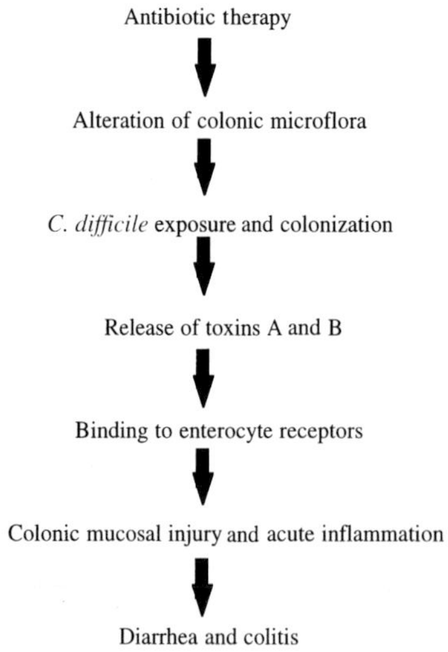

Antibiotic therapy

⬇

Alteration of colonic microflora

⬇

C. difficile exposure and colonization

⬇

Release of toxins A and B

⬇

Binding to enterocyte receptors

⬇

Colonic mucosal injury and acute inflammation

⬇

Diarrhea and colitis

Fig. 1. Pathogenesis of *Clostridium difficile*-induced diarrhea and colitis

The "big three" classes of antibiotics predisposing to *C. difficile*-induced enteric disease are cephalosporins, ampicillin/amoxicillin, and clindamycin (KELLY and LaMONT 1993). While early work focused attention on the prominent role of clindamycin as an inducing agent, most studies after 1980 show that cephalosporins have become the most common agents implicated in *C. difficile* colitis, especially in nosocomially acquired cases (GOOLEDGE et al. 1989). Ampicillin, amoxicillin, or amoxicillin-clavulanate (Augmentin) are also common causes, especially in outpatients. While the duration of antibiotic therapy and the number of different antibiotics used significantly influences the risk of *C. difficile* colitis, pseudomembranous colitis associated with a single preoperative dose of cephalosporin has been reported (FREIMAN et al. 1989). Less commonly implicated antibiotics include macrolides (erythromycin, clarithromycin, and azithromycin), tetracyclines, sulfonamides, trimethoprim, chloramphenicol, quinolines, and penicillins other than ampicillin/amoxicillin. Antibiotics that are rarely or never associated with *C. difficile* infection include parenteral aminoglycosides, vancomycin, bacitracin, nitrofurantoin, or antimicrobial agents whose activity is restricted to fungi, mycobacteria, parasites, or viruses. Curiously, even though metronidazole is considered the antibiotic of first choice in treating *C. difficile* infection, its use has occasionally been associated with induction of disease (SAGINUR et al. 1980; THOMSON et al. 1981). Antineoplastic agents have occasionally been implicated, principally methotrexate (ANAND and GLATT 1993).

Pathogenic strains of *C. difficile* produce two potent toxins, toxin A and toxin B. These high molecular weight proteins bind to specific receptors on the luminal aspect of the colonic epithelium and are then transported into the cytoplasm. Once internalized, both toxins inactivate Rho proteins, a family of small GTP-binding proteins important in actin polymerization, cytoskeletal architecture, and cell movement. The critical enzymatic action is the addition of glucose to a specific threonine on Rho (JUST et al. 1995a,b). Depolymerization of actin filaments, disruption of the cytoskeleton, cell rounding, and cell death result from Rho inactivation. Unlike cholera toxin or *Escherichia coli* heat stable toxin, *C. difficile* toxins have no effects on intracellular levels of cyclic AMP or GMP. However, recent evidence indicates that a number of other bacterial toxins target Rho proteins (VON EICHEL-STREIBER et al. 1996). For example, the cytotoxins from *C. sordellii* and *C. novyi* add a glucose to Rho and toxins from *Bacillus cereus* and *Staphylococcus aureus* also modify Rho family proteins. Thus it appears that *C. difficile* toxins and other structurally unrelated bacterial cytotoxins modify host cell structure and function by attacking Rho family proteins that are vital for maintenance of normal cell architecture.

Both *C. difficile* toxins bind to and damage human colonic epithelial cells (RIEGLER et al. 1995). In rabbit ileum, the brush border ectoenzyme, sucrase-isomaltase, binds *C. difficile* toxin A and functions as a cell surface receptor (POTHOULAKIS et al. 1996). Since this enzyme is not present in human colonic mucosa, other membrane surface glycoproteins presumably serve as toxin receptors. *C. difficile* toxins produces colonic injury in the human colon, as in cells in culture, as a result of damage to the enterocyte cytoskeleton and disruption of tight junction function (RIEGLER et al. 1995; HECHT et al. 1992). The toxins also cause a severe inflammatory reaction in the lamina propria with the formation of micro-ulcerations of the colonic mucosa that are covered by an inflammatory pseudo-membrane.

A major characteristic of *C. difficile* infection is the intense acute neutrophilic inflammation seen in pseudomembranous colitis patients and in animal models of this disease. In contrast to cholera toxin, which stimulates massive intestinal fluid secretion without a significant inflammatory response, *C. difficile* toxin A stimulates fluid secretion accompanied by considerable mucosal edema and necrosis. Al-though toxin A impairs tight junction permeability in human and animal intestine via its effects on cytoskeletal proteins, this by itself cannot explain the acute in-flammatory response observed in animal models of toxin A-mediated enteritis. Recent results provide evidence that interactions between neuropeptides and in-flammatory mediators released from inflammatory cells of the intestinal lamina propria and from epithelial cells themselves are critical initiators of the toxin A-induced inflammatory process. For example, within 15 min of luminal application of toxin A in experimental animals we observed release of the neuropeptides sub-stance P (SP) and calcitonin gene-related peptide (CGRP) from sensory nerves, and degranulation of mast cells (POTHOULAKIS et al. 1998). This is followed by release of tumor necrosis factor-α (TNF-α) from macrophages and upregulation of adhesion molecules on endothelial cells, allowing neutrophil attachment and invasion. The importance of sensory neuropeptides in this process is demonstrated by a recent report that prevention of SP and CGRP release from sensory neurons by adminis-tration of specific SP or CGRP antagonists substantially inhibit toxin-mediated diarrhea and inflammation (KEATES et al. 1998). Moreover, mice genetically defi-cient in the NK-1 (SP) receptor are largely protected from the secretory and in-flammatory changes induced by toxin A and mast cell-deficient mice have markedly diminished responses to the toxin (WERSHIL et al. 1998).

How does toxin A in the intestinal lumen initiate this early interaction between sensory nerves and mast cells and macrophages of the intestinal lamina propria? Intestinal epithelial cells release the proinflammatory chemokine macrophage in-flammatory protein-2 (MIP-2) within 15 min of exposure to toxin A, well before the onset of fluid secretion of inflammation (CASTGLIUOLO et al. 1998). Moreover, an antibody to MIP-2 substantially inhibited intestinal secretion and inflammation in this model, supporting the view that release of this chemokine is critical for pathogenesis. These results suggest that inflammatory mediators such as MIP-2 and interleukin-1β (IL-1β) released from enterocytes in response to toxin A activate sensory nerves in the subjacent lamina propria. Sensory nerves then release pro-inflammatory neuropeptides such as substance P and CGRP. These substances in turn stimulate inflammatory cells leading to release of proinflammatory cytokines, such as TNF-α and leukotrienes, that elicit neutrophil recruitment via activation of adhesion molecules on vascular endothelial cells. Indeed, pretreatment of rabbits with a monoclonal antibody directed against the neutrophil adhesion molecules CD18 substantially reduced toxin A-induced secretion and inflammation (KELLY et al. 1994a).

3 Pathology

When the human colon is exposed to *C. difficile* toxins, loss of actin filaments leads to cell rounding and shedding of cells from the basement membrane into the lumen, leaving a shallow ulcer on the mucosal surface. Serum proteins, mucus, and inflammatory cells flow outward from the ulcer, creating the typical colonic pseudomembrane. The spewing forth of the inflammatory exudate from the mucosal microulceration produces the typical "volcano" or "summit" lesion of *C. difficile* colitis (Fig. 2). On gross or sigmoidoscopic inspection of the colonic or rectal mucosa, pseudomembranes appear as yellow or off-white raised plaques 0.2–2.0cm in diameter scattered over a fairly normal-appearing intervening mucosa (Fig. 3). Edema and hyperemia of the full thickness of the bowel wall are common, and this is reflected by the typical radiographic appearance of "thumb-printing", or

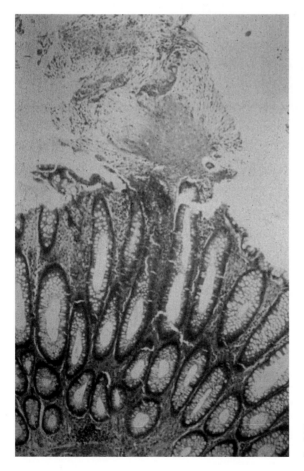

Fig. 2. A typical "volcano" or "summit" lesion of *Clostridium difficile* colitis

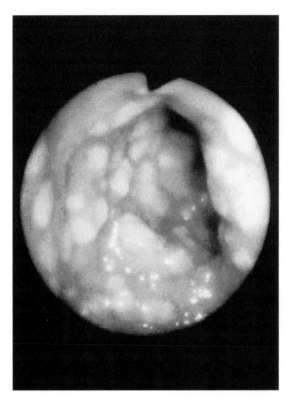

Fig. 3. Classic pseudomembranes seen on sigmoidoscopic examination

massive wall thickening on computerized tomography scanning of patients with pseudomembranous colitis.

The patchy distribution of the pseudomembranes is probably related to a toxin dose-response effect. For example, when human colonic mucosal strips in vitro were exposed to different concentrations of toxin B, cellular damage was very patchy at low concentrations. However, as the toxin concentration was raised, the area of damage increased until it was nearly confluent (RIEGLER et al. 1995). Similarly, some patients with early pseudomembranous colitis have only scattered lesions on the colonic mucosa, while others exhibit a confluent pseudomembrane covering the entire mucosa.

The pathologic features of pseudomembranous colitis (PMC) have been classified into three distinct types (PRICE and DAVIES 1977). In type 1 PMC, the mildest form, the major inflammatory changes are confined in the superficial epithelium and immediately subjacent lamina propria. Typical pseudomembranes and summit lesions are present, and crypt abscesses are occasionally noted. Type 2 PMC is characterized by more severe disruption of glands and marked mucin secretion, and more intensive inflammation of basal lamina. Type 3 PMC is characterized by severe, intense necrosis of the full thickness of the mucosa with a confluent pseudomembrane.

3.1 Immunity and Host Defenses

As noted above, the first line of defense against *C. difficile* infection is the normal bowel microflora, which inhibits growth of this pathogen in vitro and in vivo (BORRIELO 1990). Indeed, healthy adults not exposed to antibiotics are rarely infected with *C. difficile*, in contrast to healthy infants, 50% of whom are asymptomatic carriers. A second protective factor is gastric acid which reduces the number of viable spores and inactivates ingested toxins (McFARLAND et al. 1990). Normal intestinal peristalsis is also important as a defense mechanism by eliminating *C. difficile* and its toxins. Conversely, antidiarrheal medications that reduce intestinal peristaltic activity may delay clearance of the organism and worsen the duration or intensity of illness.

The humoral immune response appears to be important in determining the duration and severity of *C. difficile* colitis (KELLY 1996). Immunization of laboratory animals against toxin A protects against a subsequent challenge with *C. difficile* (KIM et al. 1998; LIBBY et al. 1982). Infant hamsters who drink milk from mothers immunized against toxins A and B are also protected (KIM et al. 1987). Although immunization of laboratory animals protects against experimental *C. difficile* colitis, it is not yet established whether circulating antitoxin A and B antibodies protect against diarrhea and colitis in humans. However, mounting evidence would appear to support a protective role for toxin antibodies in human disease. Serum antibodies against *C. difficile* toxins are present in the majority of the adult population (VISCIDI et al. 1983). Secretory IgA antitoxin is present in colonic secretions and can inhibit binding of toxin A to its specific brush border receptor providing a possible mechanism of immune protection (KELLY et al. 1992). High levels of serum and intestinal antitoxin antibodies may be associated with mild colitis or asymptomatic carriage of *C. difficile* (MULLIGAN et al. 1993). Conversely, a deficient antibody response may predispose to severe or recurrent *C. difficile* colitis (LEUNG et al. 1991; WARNY et al. 1994). Finally, treatment with pooled intravenous γ-globulin containing high levels of antibody to toxins appears to hasten recovery of patients with severe or relapsing disease (LEUNG et al. 1991; SALCEDO et al. 1997).

4 Risk Factors for Infection

C. difficile infection occurs primarily in hospitalized patients, causing as many as 3 million cases of diarrhea and colitis per year in the United States (McFARLAND et al. 1989; JOHNSON et al. 1990a), compared to only 20,000 cases per year in outpatients (HIRSCHHORN et al. 1994). A careful epidemiological study by McFARLAND and colleagues found that 20% of patients admitted to a general hospital were either culture positive on admission or acquired the organism during their hospital stay (McFARLAND et al. 1989). Interestingly, only one third of all infected patients developed diarrhea, while the remaining two thirds were asymptomatic carriers. Presumably these asymptomatic patients serve as a reservoir of infection. *C. difficile* survives in the hospital environment as antibiotic-resistant spores which are ingested by patients. The organism and its spores are detectable on toilets, bathroom floors, telephones, call buttons, stethoscopes, and the hands of health care workers (McFARLAND et al. 1989; JOHNSON et al. 1990a; FEKETY et al. 1981; KIM et al. 1981). A known risk factor for infection is sharing a hospital room with an infected patient (McFARLAND et al. 1989). Thus, cross-infection may occur by patient-to-patient spread or through environmental contamination (McFARLAND et al. 1989; JOHNSON et al. 1990a; FEKETY et al. 1981; KIM et al. 1981; NOLAN et al. 1987). The spread of infection can be interrupted by careful hand washing after examining patients and by the use of disposable gloves (JOHNSON et al. 1990b).

While antibiotic exposure is the most important risk factor for *C. difficile* infection, other risk factors include increasing age (after infancy), severity of underlying disease, nonsurgical gastrointestinal procedures, presence of a nasogastric tube, antiulcer medications, stay in an intensive care unit, and duration of hospital stay (BIGNARDI 1998) (see Table 1). An increased incidence of *C. difficile* infection in cancer and HIV patients appears to be related to specific risk factors among these groups of patients. Low intensity of chemotherapy, reflecting less frequency of neutropenia, lack of parenteral vancomycin use, and hospitalization within the previous 2 months were independently predictive of *C. difficile* colitis in hospitalized cancer patients (HORNBUCKLE et al. 1998). A CD4$^+$ cell count less than 50/mm^3 in addition to clindamycin and penicillin use were independent factors significantly associated with *C. difficile* colitis among HIV infected patients (BARBUT et al. 1997).

5 Clinical Manifestations

Infection with *C. difficile* can produce a wide spectrum of clinical manifestations ranging from the asymptomatic carrier state in infants and adults to fulminant colitis with megacolon or perforation (Table 2). The basis for this variability in

Table 1. Risk factors for *Clostridium difficile* infection

1. Hospitalization
2. Severe underlying diseases
3. Exposure to antibiotics
4. Sharing a hospital room with an infected patient
5. Increasing age
6. Nasogastric tube
7. Antiulcer medication
8. Cancer chemotherapy
9. HIV infection

response is not entirely clear, but host factors appear to be more important than bacterial virulence factors. For example, asymptomatic infant or adult carriers may be colonized with the same toxigenic strain as symptomatic patients with severe diarrhea (VISCIDI et al. 1981). Many infant carriers have very high toxin levels in their stools without diarrhea or colitis. Important host response factors may include toxin receptor density, antitoxin antibody levels, and the presence or absence of the normal barrier flora. The onset of *C. difficile* diarrhea usually occurs 4–9 days after the beginning of antibiotic therapy. However, up to one third of patients develop diarrhea after antibiotics have been discontinued, a feature that often leads to diagnostic confusion. Cancer chemotherapy may also disrupt the colonic microflora leading to *C. difficile* colonization, diarrhea, and colitis.

5.1 *Clostridium difficile* in Infants and Children

Approximately 50% of infants and children younger than 1 year commonly harbor *C. difficile* and its toxin without suffering deleterious consequences. The reason is not known. The isolation rate in infants ranges from 5% to 80%; controlled studies show that the rate of carriage is similar in children with and without enteric disease

Table 2. Clinical spectrum of *Clostridium difficile*-induced disease

Antibiotic-associated diarrhea
Mild, self-limiting diarrhea
Watery, diarrhea with mucus
No systemic symptoms
Sigmoidoscopy: normal or mild erythema, no colitis
C. difficile colitis without pseudomembrane formation
Moderate watery diarrhea, may have fecal leukocytes
Nausea, anorexia, crampy abdominal pain
Systemic symptoms including malaise, dehydration, low-grade fever
Polymorphonuclear leukocytosis
Sigmoidoscopy: diffuse/patchy erythematous colitis, no pseudomembranes
Pseudomembranous colitis
Severe, persistent watery diarrhea (15–30 stools per day)
Nausea, anorexia, crampy abdominal pain, lower quadrant tenderness
High-grade fever, severe leukocytosis ($>20,000$)
Hypoalbuminemia (<3.0g/dl), anasarca
Sigmoidoscopy: diffuse/patchy erythematous colitis, pseudomembranes
CT of abdomen: diffuse/patchy thickened colon folds ("accordion sign")
Unusual features: neutrocytic ascites, rectal bleeding, polyarthritis, ileal involvement
Fulminant colitis
Severe, watery diarrhea (may be minimal if ileus develops)
Severe lower quadrant tenderness, diffuse abdominal pain
Abdominal distension secondary to ileus, toxic megacolon, perforation
High fever, chills, dehydration, hypotension, acidosis, leukocytosis ($>30,000$mm^3)
Severe hypoalbuminemia, marked protein-losing enteropathy, ascites
Limited sigmoidoscopy: pseudomembranes (may be confluent)

(SNYDER 1940; LARSON et al. 1982; WELCH and MARKS 1982). The relatively high carriage rates persist during the first 8 months of life, until the "normal bowel flora" becomes established (BORRIELLO 1990). Children aged 2 years or older may develop antibiotic-associated pseudomembranous colitis (VISCIDI et al. 1981; KIM et al. 1983) but the incidence is less than observed in adults. Population-based studies in Sweden have shown a 20- to 100-fold increase in the incidence of *C. difficile* toxin-positive stools in people over 60 years compared to those aged 10–20 years (ARONSSON et al. 1985). Serologic assays show that most healthy children over 2 years old and adults, including those in their 70s and 80s, have circulating antibodies to toxins A and B, but whether these antibodies are protective is not established.

5.2 Adult Carrier State

The majority of hospitalized patients infected by *C. difficile* are asymptomatic carriers (McFARLAND et al. 1989), who serve as a silent reservoir for continued *C. difficile* contamination of the hospital environment. These carriers seldom develop overt or symptomatic disease, that is, the carrier state is stable. A recent longitudinal study (SHIM et al. 1998) in which 810 patients admitted to hospital were followed by prospective rectal-swab culture showed that primary asymptomatic *C. difficile* carriers have a decreased risk of *C. difficile*-associated diarrhea compared to noncolonized patients (1.0% vs 3.6%, $p = 0.02$). When those patients who received antibiotics were compared separately, the risk of *C. difficile*-associated diarrhea was further reduced among asymptomatic carriers compared to noncolonized patients (1.1% vs 4.5%, $p = 0.02$). Treatment of adult carriers with metronidazole or vancomycin is not recommended as this appears to prolong the carrier state (JOHNSON et al. 1992).

5.3 Antibiotic-Associated Diarrhea

Mild diarrhea is quite common during treatment with antibiotics, but in only 20% of cases is it related to *C. difficile*. Most antibiotic diarrhea is related to an osmotic effect of unabsorbed carbohydrate (RAO et al. 1988). In healthy individuals, unabsorbed dietary carbohydrate delivered to the large intestine undergoes fermentation by the other flora to short chain fatty acids, hydrogen, methane, and other metabolites (see Fig. 4). However, during antibiotic therapy this normal fermentation process is interrupted, allowing accumulation of carbohydrates that bind water and cause diarrhea. Diarrhea is watery, containing mucus but not blood. Sigmoidoscopic examination reveals normal colonic mucosa or mild edema or hyperemia of the rectum. Obvious colitis or pseudomembrane formation does not occur. Systemic symptoms are absent and diarrhea stops when antibiotics are discontinued in the majority of patients.

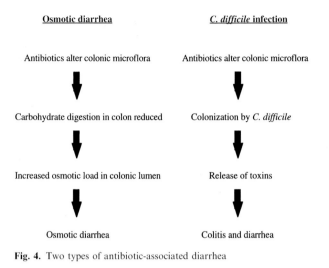

Fig. 4. Two types of antibiotic-associated diarrhea

5.4 *C. difficile* Colitis Without Pseudomembrane Formation

This is the most common clinical manifestation of *C. difficile* infection. Colitis without pseudomembranes produces a more serious illness than benign or simple antibiotic-associated diarrhea. Malaise, abdominal pain, nausea, anorexia, and watery diarrhea are present, and some patients may complain of abdominal pain and cramps that are relieved by passage of diarrhea. They may also manifest dehydration and a low-grade fever with a systemic polymorphonuclear leukocytosis. Fecal leukocytes may be present in the stools but are not a reliable indicator of colitis as they were absent in 72% of toxin positive stools in one study (MARX et al. 1993). Sigmoidoscopy may reveal a nonspecific diffuse or patchy erythematous colitis without pseudomembranes.

5.5 Pseudomembranous Colitis

This entity is the classic manifestation of full-blown *C. difficile* colitis and is accompanied by similar, but often more severe, symptoms than observed in colitis pseudomembranes. Sigmoidoscopic examination reveals the classic pseudomembranes, raised yellow plaques ranging from 2–10mm in diameter scattered over the colorectal mucosa (Fig. 3). In severely ill patients, white blood cell counts of 20,000 or greater and hypoalbuminemia of 3.0g/dl or lower may be observed. Most patients with pseudomembranous colitis have involvement of the rectosigmoid area but as many as one third of patients have pseudomembranes limited to the more proximal colon necessitating colonoscopy (TEDESCO et al. 1982). There have been a few reported cases of pseudomembrane formation involving the small intestine (TSUTAOKA et al. 1994). A number of these were in postsurgical patients and included involvement of a defunctionalized limb of a jejunal-ileal bypass (KRALOVICH et al. 1997), an ileal conduit (SHORTLAND et al. 1983), or an end-ileostomy (Fig. 5). Although an abdominal CT scan in patients with pseudomembranous colitis is not highly specific, it often reveals pronounced thickening of the colonic wall that may involve the entire colon, collections of fluid in the lower abdomen or pelvis, as well as the characteristic "accordion sign" of contrast trapped among the thickened folds (Fig. 6) (FISHMAN et al. 1991). A neutrocytic ascites with low serum to ascites albumin gradient may occur in patients with hypoalbuminemia or acquired immunodeficiency syndrome (ZUCKERMAN et al. 1997; JAFRI and MARSHALL 1996). Ascites may even be the presenting manifestation of pseudomembranous colitis. A recent radiological review of typical sonographic appearances of common colonic diseases reported ascites in 64% of patients with pseudomembranous colitis compared to 24% of patients with diverticulitis, cancer, or inflammatory or ischemic bowel disease (TRUONG et al. 1998).

5.6 Fulminant Colitis

Fulminant colitis in *C. difficile* infection occurs in approximately 3% of patients but accounts for most of the serious complications including perforation, prolonged ileus, megacolon, and death (RUBEN et al. 1995). Patients with fulminant colitis

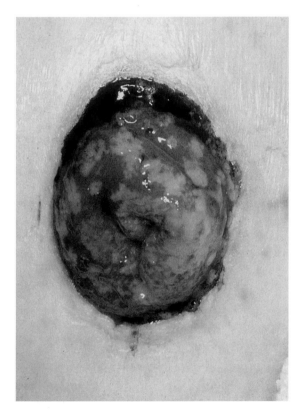

Fig. 5. Pseudomembranous enteritis involving an ileostomy site

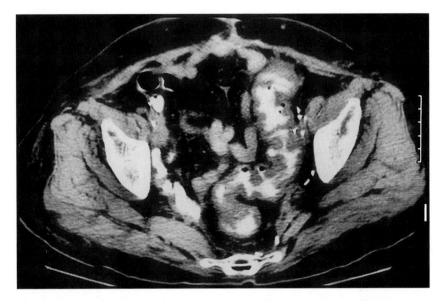

Fig. 6. A CT image of a patient abdomen with pseudomembranous colitis demonstrating the classic "accordion" sign of contrast trapped among thickened colonic folds

complain of severe lower quadrant or even diffuse abdominal pain, diarrhea, and distention. Some patients exhibit high fever, chills, and marked leukocytosis. Diarrhea is usually prominent, but may be minimal in patients who develop an ileus resulting in the pooling of secretions in the dilated, atonic colon. Hypoalbuminemia may also occur because of a severe protein-losing enteropathy. Toxic megacolon may develop and is diagnosed based on the findings of a dilated colon (> 7cm in its greatest diameter) accompanied by signs and symptoms of severe toxicity (fever, chills, dehydration, high white blood cell count). Patients with megacolon may also have dilated small intestine on plain abdominal radiographs with air–fluid levels mimicking an intestinal obstruction or ischemia (pseudo-obstruction).

In some patients, fulminant *C. difficile* infection may present with signs and symptoms of bowel perforation. Typically, these patients have abdominal rigidity, involuntary guarding, rebound tenderness, and reduced bowel sounds. Abdominal radiographs may reveal the presence of free abdominal air. Urgent surgery with partial colectomy and diverting ileostomy is required to reduce mortality and morbidity.

Aggressive diagnostic and therapeutic interventions are often necessary to prevent further morbidity or mortality in patients with fulminant *C. difficile* colitis (RUBEN et al. 1995). Limited flexible sigmoidoscopy or colonoscopy may be performed at the bedside but because of the risk of perforation only minimal amounts of air should be introduced. The presence of pseudomembranes in the rectum or sigmoid colon is sufficient to make a presumptive diagnosis of *C. difficile* colitis. A CT scan of the abdomen is also a useful diagnostic test to demonstrate colitis and to look for evidence of perforation or abscess formation. The presence of extreme leukocytosis (> 30,000), fever, hypotension, and metabolic acidosis are danger signs in fulminant colitis, and may indicate the need for laparotomy and colectomy in patients with impending or actual perforation (MORRIS et al. 1994). RAMASWAMY and colleagues determined that the following factors predicted increased mortality in severe *C. difficile* colitis: a low serum albumin on admission to hospital (< 2.5g/dl); a fall in albumin greater than 1.1g/dl at the onset of symptoms; exposure to more than three antibiotics; and persistent toxin in the stools 7 days or longer after therapy (RAMASWAMY et al. 1996).

5.7 Relapsing Infection

Approximately 15–20% of patients treated successfully for *C. difficile* diarrhea relapse following successful therapy (KELLY et al. 1994). Relapse is manifested by the abrupt reappearance of diarrhea and other symptoms usually within a week of stopping treatment with vancomycin or metronidazole. Patients who relapse once are at an even greater risk of further relapses and in one study the relapse rate in patients who had suffered two or more previous relapses was an impressive 65% (MCFARLAND et al. 1994). In addition to the number of previous *C. difficile*-associated diarrhea episodes, onset of the initial disease in spring, exposure to additional antibiotics for treatment of other infections, infection with immunoblot type 1 or 2 strains of *C. difficile*, and female gender have recently been identified by one group as factors significantly associated with high risk of *C. difficile*-associated diarrhea (FEKERTY et al. 1997).

Relapse is not related to antibiotic resistance since this has not been reported in *C. difficile*. It is not clear whether relapse is due to persistent antibiotic-resistant spores within the bowel or reinfection from the patient's environment. One suggestion is that *C. difficile* spores persist in colonic diverticula that are not in the main fecal flow, thus escaping elimination. A recent study of early recurrent *C. difficile*-associated diarrhea, defined as occurring less than 45 days after the initial episode, demonstrated identical strains of *C. difficile* from both episodes in seven out of eight cases, indicating that relapse is more common than reinfection (Do et al. 1998). It is worth emphasizing that therapy with metronidazole or vancomycin perpetuates disruption of the colonic microflora and therefore predisposes to reinfection with *C. difficile*. Possible mechanisms for relapse include reinfection from environment contamination by *C. difficile* spores, continued susceptibility to *C. difficile* colonization because of recent antibiotic therapy (with metronidazole or vancomycin), or impaired host immune responses to *C. difficile* toxins.

5.8 *C. difficile* Infection in Patients with Chronic Inflammatory Bowel Disease

Infection with *C. difficile* may complicate the course of ulcerative colitis or Crohn's disease (BOLTON et al. 1980; LAMONT and TRNKA 1980). Patients with inflammatory bowel disease (IBD) are often exposed to antibiotics and are frequently hospitalized, placing them at increased risk. IBD patients infected with *C. difficile* may manifest diarrhea, abdominal pain, and low-grade fever mimicking a flare of their IBD. Diagnosis is established by identification of *C. difficile* toxin in stool samples. Pseudomembrane formation is rare in this setting probably because the colon is already involved by inflammatory bowel disease. *C. difficile* infections respond promptly to appropriate therapy with metronidazole or vancomycin. Some patients have developed *C. difficile* at the onset of their first attack of inflammatory bowel disease, a situation that can lead to considerable diagnostic confusion. Infection with *C. difficile* in patients with ulcerative colitis or Crohn's disease requires prompt diagnosis and management since failure to diagnose the infection may lead to inappropriate treatment with corticosteroids or immunosuppressive agents.

References

Anand A, Glatt AE (1993) *Clostridium difficile* infection associated with antineoplastic chemotherapy: a review. Clin Infect Dis 17:109

Aronsson B, Molby R, Nord CE (1985) Antimicrobial agents and *Clostridium difficile* in acute disease: epidemiological data from Sweden, 1980–1982 J Infect Dis 151:476

Barbut F, Meynard JL, Guiget M, Avesani V, Bochet MV, Meyohas MC, Delmee M, Tilleul P, Frottier J, Petit JC (1997) *Clostridium difficile*-associated diarrhea in HIV-infected patients: epidemiology and risk factors. J Acquir Immune Defic Syndr Hum Retrovirol 16(3):176–181

Bartlett JG (1981) Antimicrobial agents implicated in *Clostridium difficile* toxin-associated diarrhea or colitis. Johns Hopkins Med J 149(1):6–9

Bartlett JG, Chang TW, Gurwith M, Gorbach SL, Onderdonk AD (1978) Antibiotic-associated pseudomembranous colitis due to toxin-producing clostridia. N Engl J Med 298:531–534

Bignardi GE (1998) Risk factors for *Clostridium difficile* infection. J Hosp Infect 40(1):1–15

Bolton RP, Sherriff RJ, Read AE (1980) *Clostridium difficile* associated diarrhea: a role in inflammatory bowel disease? Lancet 1:383–384

Borriello SP (1990) The influence of the normal flora on *Clostridium difficile* colonisation of the gut. Ann Med 22:61–67

Castagliuolo I, Keates AC, Wang CC, Pasha A, Valenick L, Kelly CP, Nikulasson ST, LaMont JT, Pothoulakis C (1998) *Clostridium difficile* toxin A stimulates macrophage-inflammatory protein-2 production in rat intestinal epithelial cells. J Immunol 160(12):6039–6045

Do AN, Fridkin SK, Yechouron A, Banerjee SN, Killgore GE, Bourgault AM, Jolivet M, Jarvis WR (1998) Risk factors for early recurrent *Clostridium difficile*-associated diarrhea. Clin Infect Dis 26(4):954–959

Fekety R, Kim KH, Brown D, Batts DH, Cudmore M, Silva Jr J (1981) Epidemiology of antibiotic-associated colitis; isolation of *Clostridium difficile* from the hospital environment. Am J Med 70:906–908

Fekerty R, McFarland LV, Surawicz CM, Greenberg RN, Elmer GW, Mulligan ME (1997) Recurrent *Clostridium difficile* diarrhea: characteristics of and risk factors for patients enrolled in a prospective, randomized, double-blinded trial. Clin Infect Dis 24(3):324–333

Fishman E, Kavuru M, Jones B, Kuhlman JE, Merine DS, Lillimore KD, Siegelman SS (1991) CT of pseudomembranous colitis: Radiologic, clinical, and pathologic correlation. Radiology 180:157–161

Freiman JP, Graham DJ, Green L (1989) Pseudomembranous colitis associated with single-dose cephalosporin prophylaxis. JAMA 262(7):902

Golledge CL, McKenzie T, and Riley TV (1989) Extended spectrum cephalosporins and *Clostridium difficile*. J Antimicrob Chemother 23:929–931

Hall IC, O'Toole E (1935) Intestinal flora in newborn infants with description of a new pathogenic anaerobe. Am J Dis Child 49:390–402

Hecht G, Koutsouris A, Pothoulakis C, LaMont JT, Madara JL (1992) *Clostridium difficile* toxin B disrupts the barrier function of T84 monolayers. Gastroenterology 102:416–423

Hirschhorn LR, Trnka Y, Onderdonk A, Lee ML, Platt R (1994) Epidemiology of community-acquired *Clostridium difficile*-associated diarrhea. J Infect Dis 169:127–133

Hornbuckle K, Chak A, Lazarus HM, Cooper GS, Kutteh LA, Gucalp R, Carlisle PS, Sparano J, Parker P, Salata RA (1998) Determination and validation of a predictive model for *Clostridium difficile* diarrhea in hospitalized oncology patients. Ann Oncol 9(3):307–311

Jafri SF, Marshall JB (1996) Ascites associated with antibiotic-associated pseudomembranous colitis. South Med J 89(10):1014–1017

Johnson S, Clabots CR, Linn FV, Olson MM, Peterson LR, Gerding DN (1990a) Nosocomial *Clostridium difficile* colonisation and disease. Lancet 336:97–100

Johnson S, Gerding DN, Olson MM, Weiler MD, Hughes RA, Clabots CR, Peterson LR (1990b) Prospective, controlled study of vinyl glove use to interrupt *Clostridium difficile* nosocomial transmission. Am J Med 88:137–140

Just I, Selzer J, Wilm M, von Eichel-Streiber C, Mann M, Aktories K (1995a) Glucosylation of Rho proteins by *Clostridium difficile* toxin B. Nature 375:500–503

Just I, Wilm M, Selzer J, Rex G, von Eichel-Streiber C, Mann M, Aktories K (1995b) The enterotoxin from *Clostridium difficile* (ToxA) monoglucosylates the Rho proteins. J Biol Chem 270:13932–13936

Keates AC, Castagliuolo I, Qiu B, Nikulasson S, Sengupta A, Pothoulakis C (1998) CGRP upregulation in dorsal root ganglia and ileal mucosa during *Clostridium difficile* toxin A-induced enteritis. Am J Physiol 274(1 Pt 1):G196–202

Kelly CP, Pothoulakis C, Orellana J, LaMont JT (1992) Human colonic aspirates containing immunoglobulin A antibody to *Clostridium difficile* toxin A inhibit toxin A-receptor binding. Gastroenterology 102:35–40

Kelly CP, LaMont JT (1993) Treatment of *Clostridium difficile* diarrhea and colitis. In: Wolfe, MM (ed) Gastrointestinal Pharmacotherapy. pp 199–212. Philadelphia: Saunders

Kelly CP, Becker S, Linevsky JK, Joshi MA, O'Keane JC, Dickey BF, LaMont JT, Pothoulakis C (1994a) Neutrophil recruitment in *Clostridium difficile* toxin A enteritis in the rabbit. J Clin Invest 93(3):1257–1265

Kelly CP, Pothoulakis C, LaMont JT (1994b) *Clostridium difficile* colitis. N Engl J Med 330:257–262

Kelly CP (1996) Immune response to *Clostridium difficile* infection. Eur J Gastroenterol Hepatol 8:1048–1053

Kim KH, Fekety R, Batts DH, Brown D, Cudmore M, Silva Jr J, Waters D (1981) Isolation of *Clostridium difficile* from the environment and contacts of patients with antibiotic-associated colitis. J Infect Dis 143:42–50.

Kim KH, DuPont HL, Pickering LK (1983) Outbreaks of diarrhea associated with *Clostridium difficile* and its toxin in day-care centers: Evidence of person-to-person spread. J Pediatr 102:376

Kim KH, Iaconis JP, Rolfe RD (1987) Immunization of adult hamsters against *Clostridium difficile*-associated ileocecitis and transfer of protection to infant hamsters. Infect Immun 55:2984–2992

Kralovich KA, Sacsner J, Karmy-Jones RA, Eggenberger JC (1997) Pseudomembranous colitis with associated fulminant ileitis in the defunctionalized limb of a jejunal-ileal bypass: case report. Dis Colon Rectum 40(5):622–4

LaMont JT, Trnka YM (1980) Therapeutic implications of *Clostridium difficile* toxin during relapse of chronic inflammatory bowel disease. Lancet 1:381–383

Larson HE, Barclay FE, Honour P, Hill ID (1982) Epidemiology of *Clostridium difficile* in infants. J Infect Dis 146:727

Leung DY, Kelly CP, Boguniewicz M, Pothoulakis C, LaMont JT, Flores A (1991) Treatment with intravenously administered gamma globulin of chronic relapsing colitis induced by *Clostridium difficile* toxin. J Pediatr 118:633–637

Libby JM, Jortner BS, Wilkins TD (1982) Effects of the two toxins of *Clostridium difficile* in antibiotic-associated cecitis in hamsters. Infect Immun 36:822–829

Marx CE, Morris A, Wilson ML, Reller LB (1993) Fecal leukocytes in stool specimens submitted for *Clostridium difficile* toxin assay. Diagn Microbiol Infect Dis 16(4):313–315

McFarland LV, Mulligan ME, Kwok RY, Stamm WE (1989) Nosocomial acquisition of Clostridium difficile infection. N Engl J Med 320:204–210

McFarland LV, Surawicz CM, Stamm WE (1990) Risk factors for *Clostridium difficile* carriage and *C. difficile*-associated diarrhea in a cohort of hospitalized patients. J Infect Dis 162:678–684

McFarland LV, Surawicz CM, Greenberg RN, Fekety R, Elmer GW, Moyer KA, Melcher SA, Bowen SE, Cox JL, Noorani Z (1994) A randomized placebo-controlled trial of *Saccharomyces boulardii* in combination with standard antibiotics for *Clostridium difficile* disease. JAMA 271:1913–1918

Morris LL, Villalba MR, Glover JL (1994) Management of pseudomembranous colitis. Am Surg 60: 548–552

Mulligan ME, Miller SD, McFarland LV, Fung HC, Kwok RY (1993) Elevated levels of serum immunoglobulins in asymptomatic carriers of *Clostridium difficile*. Clin Infect Dis 16 Suppl 4:S239–244

Nolan NP, Kelly CP, Humphreys JF, Cooney C, O'Connor R, Walsh TN, Weir DG, O'Briain DS (1987) An epidemic of pseudomembranous colitis: importance of person to person spread. Gut 28:1467–1473

Pothoulakis C (1996) Pathogenesis of *Clostridium difficile*-associated diarrhoea. Eur J Gastroenterol Hepatol 8:1041–1047

Pothoulakis C, Castagliuolo I, LaMont JT (1998) Neurons and mast cells modulate secretory and inflammatory responses to enterotoxins. News in Physiological Sciences 13:58–63

Price AB, Davies DR (1977) Pseudomembranous colitis. J Clin Pathol 30:1–12

Ramaswamy R, Grover H, Corpuz M, Daniels P, Pitchumoni CS (1996) Prognostic criteria in *Clostridium difficile* colitis. Am J Gastroenterol 91(3):460–464

Rao SS, Edwards CA, Austen CJ, Bruce C, Reed NW (1988) Impaired colonic fermentation of carbohydrates after ampicillin. Gastroenterology 94(4):928–932

Riegler M, Sedivy R, Pothoulakis C, Hamilton G, Zacherl J, Bischof G, Cosentini E, Feil W, Schiessel R, LaMont JT (1995) *Clostridium difficile* toxin B is more potent than toxin A in damaging human colonic epithelium in vitro. J Clin Invest 95:2004–2011

Rubin MS, Bodenstein LE, Kent KC (1995) Severe *Clostridium difficile* colitis. Dis Colon Rectum 38: 350–354

Saginur R, Hawley CR, Bartlett JG (1980) Colitis associated with metronidazole therapy. J Infect Dis 141:772–774

Salcedo J, Keates S, Pothoulakis C, Castagliuolo I, LaMont JT, Kelly CP (1997) Intravenous immunoglobulin therapy for severe *Clostridium difficile* colitis. Gut 41(3):366–370

Shim JK, Johnson S, Samore MH, Bliss DZ, Gerding DN (1998) Primary symptomless colonisation by *Clostridium difficile* and decreased risk of subsequent diarrhea. Lancet 351(9103):633–6

Shortland JR, Spencer RC, Williams JL (1983) Pseudomembranous colitis associated with changes in an ileal conduit. J Clin Pathol 36(10):1184–7

Snyder ML (1940) The normal fecal flora between two weeks and one year of age: serial studies. J Infect Dis 65:1

Tedesco FJ, Barton RW, Alpers DH (1974) Clindamycin-associated colitis. A prospective study. Ann Intern Med 81:429–433

Tedesco FJ, Corless JK, Brownstein RE (1982) Rectal sparing in antibiotic-associated pseudomembranous colitis: a prospective study. Gastroenterology 83:1259–1260

Thomson G, Clark AH, Hare K, Spilg WG (1981) Pseudomembranous colitis after treatment with metronidazole. Br Med J 282:864–865

Truong M, Atri M, Bret PM, Reinhold C, Kintzen G, Thibodeau M, Aldis AE, Chang Y (1998) Sonographic appearances of benign and malignant conditions of the colon. Am J Roentgenol 170(6):1451–1455

Tsutaoka B, Hansen J, Johnson D, Holodniy M (1994) Antibiotic-associated pseudomembranous enteritis due to *Clostridium difficile*. Clin Infect Dis 18(6):982–984

Tvede M, Rask-Madsen J (1989) Bacteriotherapy for chronic relapsing *Clostridium difficile* diarrhoea in six patients. Lancet 1:1156–1160

Viscidi R, Bartlett JG (1981) Antibiotic-associated pseudomembranous colitis in children. Pediatrics 67:381

Viscidi R, Willey S, Bartlett JG (1981) Isolation rates and toxigenic potential of *Clostridium difficile* isolates from various patient populations. Gastroenterology 81:5–9

Viscidi R, Laughon BE, Yolken R, Bo-Linn P, Moench T, Ryder RW, Bartlett JG (1983) Serum antibody response to toxins A and B of *Clostridium difficile*. J Infect Dis 148:93–100

Von Eichel-Streiber C, Boquet P, Souerborn M, Thalestam M (1996) Large clostridial cytotoxins-a family of glycosyltransferases modifying small GTP-binding proteins. Trnds Microbiol 4(10):375–382

Warny M, Vaerman JP, Avesani V, Delmee M (1994) Human antibody response to *Clostridium difficile* toxin A in relation to clinical course of infection. Infect Immun 62:384–389

Welch DF, Marks MI (1982) Is *Clostridium difficile* pathogenic in infants? J Paediatrics 100:392

Wershil BK, Castagliuolo I, Pothoulakis C (1998) Direct evidence of mast cell involvement in *Clostridium difficile* toxin A-induced enteritis in mice. Gastroenterology 114(5):956–964

Zuckerman E, Kanel G, Ha C, Kahn J, Gottesman BS, Korula J (1997) Low albumin gradient ascites complicating severe pseudomembranous colitis. Gastroenterology 112(3):991–994

Treatment of *Clostridium difficile*-Associated Diarrhea and Colitis

D.N. GERDING

1 Introduction

Treatment of *Clostridium difficile* diarrhea is well established and has not changed markedly over the past two decades (PETERSON and GERDING 1990). The major shift in therapeutic choice has been toward greater use of metronidazole rather than vancomycin because of increasing concern about vancomycin resistance in organisms such as the enterococcus (HICPAC 1995). However, this should not be interpreted to mean that improvements in therapy are not needed. There continue to be major deficiencies related to high recurrence rates following treatment, and there are no clearly effective treatments for patients with complications such as ileus, toxic megacolon, and perforation. There is increasing interest in biotherapeutic or probiotic approaches to the management of *C. difficile* disease, both for prevention and treatment (ELMER et al. 1996). Currently, evidence of efficacy for biologic agents is available only for the yeast *Saccharomyces boulardii* as an adjunctive therapy for the treatment of recurrent *C. difficile* diarrhea (McFARLAND et al. 1994).

Department of Medicine, Northwestern University Medical School, Medical Service, Veterans Affairs Chicago Healthcare System – Lakeside Division, 333 East Huron Street, Chicago, IL 60611, USA

2 Current Therapy

The common view of treatment of C. *difficile*-associated diarrhea is that metronidazole or vancomycin should be used. However, prior to a decision to initiate specific therapy it is important to note that C. *difficile*-associated diarrhea will resolve in 15–23% of patients within 2–3 days of discontinuing the offending antimicrobial treatment (TEASLEY et al. 1983; OLSON et al. 1994). This should not be construed to imply that placebo treatment is as effective as specific treatment with vancomycin or metronidazole (KEIGHLEY et al. 1978), but rather that in a subset of patients, diarrhea will resolve without specific therapy within 2–3 days of stopping the offending antibiotic. Another reason to consider withholding specific anti-C. *difficile* treatment is that there is a 5–42% risk of recurrence following specific treatment completion (ZIMMERMAN et al. 1997). It is generally assumed that there is no risk of recurrence with simple cessation of the offending antibiotic, but available studies have not recorded actual recurrence rates in this setting (TEASLEY et al. 1983; OLSON et al. 1994).

A number of general treatment guidelines for C. *difficile*-associated diarrhea have been proposed (GERDING et al. 1995; FEKETY 1997). If possible, the offending antimicrobials should be discontinued. If discontinuation is not possible, a substitution of antimicrobials that are less predisposing to C. *difficile* diarrhea (e.g., metronidazole, vancomycin, or an aminoglycoside) may be beneficial, although no specific data are available regarding benefit. Supportive measures such as fluids and electrolytes should be given to maintain hydration. Antiperistaltic agents and opiates should probably be avoided to prevent masking of symptoms and possible worsening of the disease by retaining toxin-laden secretions in the colon (NOVAK et al. 1976). Antiperistaltic agents theoretically could also lead to increased absorption of metronidazole by reducing diarrhea, and potentially cause failure of metronidazole treatment. Despite these cautions, one retrospective study failed to show any impaired response when antimotility drugs were used in conjunction with vancomycin or metronidazole for disease of mild to moderate severity (WILCOX and HOWE 1995). Finally, test-of-cure cultures or toxin assays following treatment are not recommended as they are imperfect predictors of subsequent relapse (FINEGOLD and GEORGE 1988) and treatment of asymptomatic patients colonized with C. *difficile* has been ineffective (JOHNSON et al. 1992).

2.1 Specific Treatment of Uncomplicated Initial Episodes

The subject of C. *difficile* treatment has been studied frequently in the past two decades (Table 1). Considering only prospective randomized clinical trials in humans, there have been nine studies published that have evaluated vancomycin, metronidazole, placebo, colestipol, bacitracin, teicoplanin, and fusidic acid as oral treatments (ZIMMERMAN et al. 1997). Most studies have been small; the largest single treatment arm in any trial had 52 patients treated with vancomycin vs 42

Table 1. Summary of randomized, comparative trials of oral therapy for *Clostridium difficile*-associated diarrhea (adapted from PETERSON and GERDING 1990)

Antibiotic	Regimen	Number of patients	Resolution of diarrhea (%)	Recurrence (%)	Mean days to resolution	Reference
Metronidazole	250 mg q.i.d. × 10 days	42	40 (95)	2 (5)	2.4	TEASLEY et al. 1983
	500 mg t.i.d. × 10 days	31	29 (94)	5 (17)	3.2	WENISCH et al. 1996
Vancomycin	500 mg t.i.d. × 10 days	31	29 (94)	5 (17)	3.1	WENISCH et al. 1996
	500 mg q.i.d. × 10 days	87	87 (100)	13 (15)	2.6–3.6	TEASLEY et al. 1983 DELALLA et al. 1992 DUDLEY et al. 1986
	125 mg q.i.d. × 7 days	21	18 (86)	6 (33)	4.2	YOUNG et al. 1985
	125 mg q.i.d. × 5 days	12	9 (75)	?	< 5	KEIGHLEY et al. 1978
Teicoplanin	400 mg b.i.d. × 10 days	28	27 (96)	2 (7)	2.8	WENISCH et al. 1996
	100 mg b.i.d. × 10 days	26	25 (96)	2 (8)	3.4	DELALLA et al. 1992
Fusidic acid	500 mg t.i.d. × 10 days	29	27 (93)	8 (28)	3.8	WENISCH et al. 1996
Bacitracin	25,000 U q.i.d. × 10 days	15	12 (80)	5 (42)	3.0	DUDLEY et al. 1986
	20,000 U q.i.d. × 7 days	21	16 (76)	5 (31)	4.1	YOUNG et al. 1985
Colestipol	10 gm q.i.d.	12	3 (25)	?	< 5	MOGG et al. 1982
Placebo		14	3 (21)	?	< 5	MOGG et al. 1982

treated with metronidazole (TEASLEY et al. 1983). The primary end point variable for evaluation has been clinical cessation of diarrhea. Vancomycin has been shown to be superior to placebo, whereas colestipol has been shown to be no better than placebo (KEIGHLEY et al. 1978; MOGG et al. 1982). Because of the small numbers of patients treated in each of the trials and the comparable treatment outcomes, there have been no studies that showed a statistically significant superiority of one antibiotic treatment over another. Similarly, no trials have been sufficiently large to show that regimens are statistically equivalent. Pooling of data from 469 patients studied in the randomized trials (a statistically questionable exercise because of the different protocols) also did not show a superiority of one treatment agent over another. Secondary end points evaluated have been: (a) clinical recurrence of disease, and (b) clearance of *C. difficile* or its toxin from stools following treatment. Only one randomized trial showed a significant reduction in a secondary end point (recurrence rate) when compared to the other treatment agents: teicoplanin (7%) was superior to fusidic acid (28%, $p = 0.042$) (WENISCH et al. 1996).

Despite the absence of sufficient prospective randomized studies to recommend one specific therapy over another, some principles of treatment have been derived from prior experience. When required, specific therapy should be administered orally, particularly in the case of vancomycin. Although all proven therapies have been with oral regimens, there is anecdotal experience using intravenous metronidazole in the treatment of *C. difficile*-associated diarrhea (KLEINFELD et al. 1988). Bactericidal fecal concentrations can be achieved in patients with acute disease when metronidazole is administered by the intravenous route (BOLTON et al. 1986), but in the presence of adynamic ileus, intravenous metronidazole has also been anecdotally reported to fail in the treatment of pseudomembranous colitis (GUZMAN et al. 1988). Most of the experience on specific treatment of *C. difficile* diarrhea has been obtained with either metronidazole or with vancomycin.

It is expected that nearly all patients will respond to specific therapy with vancomycin or metronidazole even after an initial recurrence following treatment with the same drug (OLSON et al. 1994). Most patients with *C. difficile*-associated diarrhea show some improvement within the first 2 days of initiating treatment, but the mean time to resolution of diarrhea ranges from 2 to 4 days in prospective studies (TEASLEY et al. 1983; WENISCH et al. 1996; PETERSON and GERDING 1990). One retrospective study (WILCOX and HOWE 1995) found the resolution time of diarrhea to be longer for metronidazole (4.6 days) than for vancomycin (3.0 days, $p < 0.01$), but this has not been confirmed in any of the prospective trials (Table 1). Patients should not be deemed therapeutic failures until at least 6 days of treatment have been allowed (PETERSON and GERDING 1990). With vancomycin treatment approximately 90% of patients respond within 7 days and 10% thereafter (FEKETY et al. 1989). Treatment is more likely to be successful if continued for 10 days. No controlled comparisons are available, but vancomycin given at a dosage of 125mg four times daily for 5–7 days (Table 1) appears to be less efficacious than when given for at least 10 days (GERDING et al. 1995). Comparable data on duration of therapy with metronidazole are not available. Most trials have used a 10-day course of metronidazole with good results (TEASLEY et al. 1983; WENISCH et al. 1996).

Regimens other than metronidazole or vancomycin that have been compared in randomized therapeutic trials for *C. difficile* diarrhea include bacitracin, teicoplanin, fusidic acid, and colestipol (Table 1). Although not statistically significant, the clinical response rates for *C. difficile* diarrhea treated with bacitracin at 20,000U or 25,000U four times per day are 10–20% lower than with vancomycin (YOUNG et al. 1985; DUDLEY et al. 1986). Bacitracin, which is not clinically readily available for use as oral treatment in the United States, should be considered as a third- or fourth-line agent in the treatment of *C. difficile* diarrhea. Treatment of 26 patients with teicoplanin at 100mg twice daily for 10 days achieved response rates (96%) similar to those achieved with vancomycin at a dose of 500mg four times per day (DELALLA et al. 1992). Teicoplanin for 10 days at a dose of 400mg twice daily in 28 patients has also been compared to vancomycin or metronidazole, each at 500mg three times daily, with equivalent response rates of 96%, 94%, and 94%, respectively (WENISCH et al. 1996). The dosing frequency of teicoplanin was found to be an important determinant of response (Swedish CDAD study group 1994). Patients randomized to receive 50mg teicoplanin four times per day for 3 days followed by 100mg twice a day for 4 additional days responded significantly better (96%) than patients randomized to receive 100mg twice a day for 7 days (70%, $p = 0.02$).

Fusidic acid, 500mg three times a day for 10 days in 29 patients, demonstrated an equivalent clinical response (93%) to that found for vancomycin, metronidazole, or teicoplanin (WENISCH et al. 1996). Although effective, neither fusidic acid nor teicoplanin are available clinically in the United States. Treatment with the ion-exchange resin colestipol is similar to placebo and clearly inferior to the antimicrobial treatment agents and is not recommended for initial treatment of *C. difficile* diarrhea (MOGG et al. 1982).

Important theoretical considerations for effective specific *C. difficile* therapy include in vitro susceptibility of *C. difficile* to the antimicrobial (Table 2) and the antimicrobial concentration at the site of infection. Summarizing data from multiple studies, the minimal inhibitory concentration for 90% of isolates (MIC_{90}) of *C. difficile* was 0.4mg/l for metronidazole, 1.6mg/l for vancomycin, 0.1mg/l for teicoplanin, and 32mg/l for bacitracin (PETERSON and GERDING 1990). Resistance to either metronidazole or vancomycin is very rare, although among 105 isolates of *C. difficile* obtained from horses, the MIC_{90} was 16mg/l for metronidazole (JANG et al. 1997). Typing of the isolates by arbitrary primed polymerase chain reaction (PCR) indicated that all but one of the isolates were from a single clone, suggesting a common source outbreak (JANG et al. 1997). Among human isolates, susceptibility of ten organisms obtained from patients who failed metronidazole therapy was compared to the susceptibility of 20 contemporary isolates of patients who responded to metronidazole. No difference in mean MIC was found between the two groups, and both were highly susceptible to metronidazole with a mean MIC of 0.23mg/l for organisms of patients that failed and 0.29mg/l for organisms of successfully treated patients (SANCHEZ et al. 1999).

In contrast to metronidazole, fecal drug concentrations of vancomycin (a relatively nonabsorbable agent) are in the range of 2000–5000mg/l (3 to 4 log_{10} higher than the MIC for *C. difficile*) (SILVA et al. 1981; JOHNSON et al. 1992). Metronidazole is well absorbed and fecal concentrations are low or absent in healthy volunteers and asymptomatic *C. difficile* carriers (HOVERSTAD et al. 1986; ARABI et al. 1979; JOHNSON et al. 1992). In contrast, bactericidal fecal concentrations were detected in all of the acute specimens obtained from nine patients with active *C. difficile* diarrhea with a mean ($\pm$ SD) concentration of 9.3 $\pm$ 7.5mg/kg stool (wet weight) (BOLTON and CULSHAW 1986). As diarrhea improved, metronidazole fecal concentrations decreased, suggesting either that the drug may be

Table 2. In vitro susceptibility of *Clostridium difficile* to therapeutic antimicrobial agents

Agent	Number of isolates	MIC (mg/l)			Reference
		MIC_{50}	MIC_{90}	Range	
Metronidazole	50	0.29	0.6	0.25–1.0	BANNATYNE et al. 1987
Vancomycin	70	1.0	2.0	1.0–2.0	BIAVSCO et al. 1991
Teicoplanin	70	0.5	0.5	0.25–1.0	BIAVSCO et al. 1991
Bacitracin	34	> 16.0	32.0	16.0–32.0	BACON et al. 1991
Fusidic acid	20	(17/20 susceptible to 10µg, disc testing)			BURDON et al. 1979
Rifampin	55	< 0.001	0.002	< 0.001–0.002	BACON et al. 1991

secreted directly through inflamed colonic mucosa during active colitis, or that more rapid intestinal transit time with diarrhea causes absorption to be decreased (INGS et al. 1975).

Metronidazole is a proven clinically effective treatment for *C. difficile* despite theoretical concerns about achieving significant fecal levels. A clinical cure rate of 98% has been compiled for metronidazole as reported for 632 patients treated in one hospital (OLSON et al. 1994). In this uncontrolled setting, the drug intolerance rate, treatment failure rate, and recurrence rate were 1%, 2%, and 7%, respectively. Metronidazole is also the least expensive treatment agent for *C. difficile* diarrhea, with a pharmacy cost of US$35 for a 10-day treatment course (500mg three times daily) compared to $287 for fusidic acid (500mg three times daily), $2030 for vancomycin (500mg three times daily), and $3340 for teicoplanin (400mg twice daily) (WENISCH et al. 1996). Lower doses of vancomycin (125mg four times daily) and teicoplanin (100mg twice daily) have similar clinical response rates at considerable cost savings compared to the doses quoted above (ZIMMERMAN et al. 1997). Metronidazole is efficacious, inexpensive, and the preferred treatment for *C. difficile*, although it does not have the United States Food and Drug Administration (FDA) approval for this use. In addition, the safety of metronidazole in children has not been proven and it is a category B drug for use in pregnancy, factors which should be taken into account when selecting therapies for these patients (FEKETY 1997).

Vancomycin was the first agent demonstrated to be highly effective for *C. difficile* diarrhea and colitis and is the drug to which all subsequent therapies have been compared (FEKETY and SHAH 1993). Despite remarkable efficacy, the increasing threat and reality of vancomycin resistance among enterococci and staphylococci in the United States has resulted in the Hospital Infection Control Practices Advisory Committee (HICPAC) for Preventing the Spread of Vancomycin Resistance making recommendations which have discouraged the use of oral vancomycin for the treatment of *C. difficile* diarrhea, except for failures of metronidazole or severe potentially life-threatening illnesses (HICPAC 1995).

2.2 Treatment of Recurrences

Perhaps the most glaring inadequacy of current therapy for *C. difficile* diarrhea is the problem of recurrence of diarrhea symptoms following successful treatment. Most recurrences resolve with re-treatment, but occasionally a patient may have numerous recurrences, the anecdotal record for which seems to be 26 such events in one patient over a period of 2.5 years (GORBACH 1996). Recurrence can be caused by the original organism (a relapse) or by a new infecting strain (a reinfection), and occurs in 5%–42% of patients successfully treated with any of the regimens employed in controlled trials (ZIMMERMAN et al. 1997). Data from three studies suggest that about half of all recurrences are caused by relapse and half by reinfection (JOHNSON et al. 1989; O'NEILL et al. 1991; WILCOX et al. 1998). It is generally accepted that the reason patients are susceptible to reinfection is that the normal bowel flora remains disrupted following treatment with metronidazole or vanco-mycin, in essence the same circumstance that predisposed the patient to the original episode of diarrhea. For relapses, the mechanism of susceptibility may differ for vancomycin and metronidazole, although this is speculative. Metronidazole fecal levels decline with treatment as diarrhea resolves, which could permit germination of any remaining *C. difficile* spores and survival of vegetative cells (BOLTON and CULSHAW 1986). In contrast to metronidazole, vancomycin has a bacteriostatic rather than bactericidal effect on *C. difficile* in vitro at the high fecal concentrations achieved during therapy, a circumstance which could permit vegetative cells to remain viable (LEVETT 1991).

It is reassuring that regardless of the mechanism of diarrhea recurrence, patients respond again to the same specific therapy, and in our experience, 92% did not have further recurrences (OLSON et al. 1994). Thus, routine use of vancomycin for second episodes of *C. difficile* diarrhea does not appear to be justified, given the concerns about vancomycin resistance and the higher cost compared to metro-nidazole (WILCOX 1998). Repeated treatment with metronidazole for first recur-rences of *C. difficile* diarrhea is our routine practice.

For patients who experience two or more recurrences, the available methods are largely anecdotal in terms of efficacy. The rationale behind some of these strategies is an attempt to reestablish the normal colonic flora, particularly treat-ment with vancomycin or metronidazole followed by the yeast *Saccharomyces boulardii* (SURAWICZ et al. 1989; MCFARLAND et al. 1994); metronidazole or bacitracin followed by *Lactobacillus* GG (GORBACH et al. 1987); vancomycin followed by synthetic fecal bacterial enema (TVEDE and RASK-MADSEN 1989); and administration of a nontoxigenic *C. difficile* strain (SEAL et al. 1987). The only randomized controlled study of treatment of *C. difficile* recurrences has been with the use of metronidazole or vancomycin standard therapy with or without the addition of *S. boulardii* for 30 days (MCFARLAND et al. 1994). The addition of *S. boulardii* significantly reduced the recurrence rate in patients with recurrent *C. difficile* diarrhea ($p = 0.04$), but did not reduce recurrences in patients who were having their first episode of *C. difficile* diarrhea ($p = 0.86$). Caution in the use of *S. boulardii* in debilitated or immunocompromised patients is recommended as invasive fungal infections and fungemia have been reported following its use in these patients (BASSETTI et al. 1998). *S. boulardii* is not commercially available in the United States and does not have FDA approval for use. Other biotherapeutic approaches with suggested benefit have been tested only in uncontrolled open protocols or reported anecdotally in small numbers of patients. These include treatment with *Lactobacillus* GG (GORBACH et al. 1987; BILLER et al. 1995), rectal infusion of feces from normal hosts (BOWDEN et al. 1981; SCHWAN et al. 1984), or infusion of a mixture of bacteria simulating fecal flora (TVEDE and RASK-MADSEN 1989). The aesthetics and safety of the bacterial rectal infusion methods pose serious impediments to their use as treatments. Oral administration of a non-toxigenic strain of *C. difficile* has been reported in two patients with partial success (one patient had a recurrence), but has not been further pursued clinically (SEAL et al. 1987).

Other uncontrolled reported approaches to patients with multiple recurrences include: treatment with vancomycin in tapering doses over 21 days followed by pulse dosing for 21 days (TEDESCO et al. 1985); no treatment with careful observation (BARTLETT 1992); vancomycin followed by the anion-exchange binding resin cholestyramine (PRUKSANANONDA and POWELL 1989, MONCINO and FALLETA 1992); and combined treatment with vancomycin and rifampin for 7 days (BUGGY et al. 1987). Although not studied in a controlled trial, use of combined vancomycin (125mg orally q.i.d.) and rifampin (600mg orally b.i.d.) is our first choice (and one of the simplest approaches) for treatment of multiple recurrences; if patients can tolerate the regimen we administer it for 10 days.

In children with chronic relapsing *C. difficile* diarrhea and low levels of serum IgG antibodies to *C. difficile* toxin A, treatment with intravenous γ-globulin has resulted in clinical and bacteriological improvement (LEUNG et al. 1991; WARNY et al. 1995). These patients have been found to have low levels of serum IgG antibody to toxin A or have had selective IgG1 deficiency. The high cost and limited availability of intravenous immunoglobulin makes this approach impractical for all but the few chronically infected children who have antibody deficiency. Whole bowel irrigation with polyethylene glycol solution followed by vancomycin treatment was successful in treating two children (LIACOURAS and PICCOLI 1996). All of these approaches to the treatment of recurrent *C. difficile* disease have practical limitations and review of the primary reference before employing any of these strategies is advised.

2.3 Medical Treatment of Complications

The most serious and controversial issue in the treatment of *C difficile* infection is the management of the rare patient who presents with or develops an acute abdomen, toxic megacolon, or ileus. These patients may have atypical symptoms, often presenting without diarrhea and mimicking an acute surgical abdomen (MORRIS et al. 1990; TRIADAFILOPOULOS and HALLSTONE 1991; TRUDEL et al. 1995). In addition, patients have been increasingly described with sepsis syndrome as a complication of severe *C. difficile* disease (CHATILLA and MANTHOUS 1995; LOWENKRON et al. 1996). Acute abdomen can occur with or without toxic megacolon, demonstrating signs of obstruction, ileus, bowel wall thickening and ascites on radiographic examination (especially CT), and often accompanied by a marked blood leukocytosis (TRIADAFILOPOULOS and HALLSTONE 1991; CHATILA and MANTHOUS 1995; LIPSETT et al. 1994). The ascitic fluid has been found to have a high white cell count, predominance of neutrophils, and a low albumin, thus mimicking spontaneous bacterial peritonitis, but cultures have been negative (ZUCKERMAN et al. 1997). The first major treatment impediment is the failure to include *C. difficile* disease as a possible cause of the atypical presentation. Delay in recognition or failure to consider *C. difficile* in this setting can lead to treatment delays and severe complications. It is imperative that *C. difficile* be included in the differential diagnosis of patients with an acute abdomen, sepsis, or toxic megacolon if the patients have received antibiotics within the previous month or two, whether diarrhea is present or not. If the diagnosis of *C. difficile* disease is considered, rapid diagnosis can best be achieved by cautious sigmoidoscopy or colonoscopy to document the presence of pseudomembranes. Overall, the likelihood of seeing pseudomembranes by endoscopy in any patient with *C. difficile* diarrhea is about 50%, so failure to visualize pseudomembranes does not rule out the possibility of *C. difficile* diarrhea (GERDING et al. 1986). A stool specimen should immediately be sent for *C. difficile* toxin testing using enzyme immunoassay as an additional rapid diagnostic measure (ALTAIE et al. 1994; ARROW et al. 1994).

Not only is diagnosis difficult, but there is no established definitive treatment for these seriously ill patients. Medical management is usually attempted if the patient is hemodynamically stable and does not have acute peritonitis. The challenge in treatment is to achieve effective antimicrobial concentrations at the site of infection (the colon) when the oral route is potentially compromised due to ileus. One experienced author advocates treatment with intravenous metronidazole or with intravenous vancomycin at dosages ≥2g/day, instillation of vancomycin via placement of a long catheter in the small intestine, or instillation of vancomycin by enema (SILVA 1989). Another similar approach (Table 3) that we have used successfully in six of eight patients with severe ileus employs vancomycin administered by nasogastric tube and by retention enema plus intravenous metronidazole (OLSON et al. 1994). None of these medical approaches has been systematically evaluated for comparative efficacy in a controlled manner.

2.4 Surgical Treatment of Complications

Patients with toxic megacolon or acute abdomen who do not respond to medical management or who have suspected colonic perforation are candidates for surgical intervention. The need for surgery in treating *C. difficile* disease is rare. LIPSETT et al. (1994) reported that 0.39% of all patients with *C. difficile* disease in one hospital required surgical intervention. However, among critically ill patients in an intensive care unit, 20.3% of patients who had *C. difficile* colitis required surgical intervention (GRUNDFEST-BRONIATOWSKI et al. 1996). The role of surgery in

Table 3. Treatment protocol for *Clostridium difficile*-infected patients with severe ileus (from OLSON et al. 1994)

Vancomycin	500mg per rectum every 6h[a]
Vancomycin	500mg via NG tube every 6h[b]
Metronidazole	500mg IV every 6h

NG, nasogastric
[a] Liquid intravenous formulation diluted in 100cc of normal saline: Insert #18 Foley catheter into rectum 4–8 in, fill the balloon to 30cc, gently pull catheter down, instill vancomycin and clamp catheter for 60min, deflate balloon and remove catheter
[b] Liquid intravenous formulation diluted with at least 10cc of fluid: Clamp nasogastric tube for 60min after each instillation

C. difficile colitis has been reviewed extensively (BRADBURY and BARRETT 1996). Mortality is high in patients who require surgery, ranging from 32% to 50%, and is predictable from the APACHE scores of the patients (PANIS et al. 1992; LIPSETT et al. 1994; MORRIS et al. 1990). LIPSETT et al. (1994) reported a lower mortality (14%) for patients undergoing subtotal colectomy (compared to 100% in patients who underwent left hemicolectomy) in an uncontrolled setting. They recommend subtotal colectomy as the procedure of choice regardless of the external appearance of the colon at surgery.

3 Summary

Treatment of *C. difficile* diarrhea with metronidazole or vancomycin is highly effective at relieving symptoms. The high rate of diarrhea recurrence is concerning, but fortunately most patients respond to a second course of treatment. The problem of vancomycin resistance in hospital organisms has markedly reduced usage of this agent as a first-line treatment for *C. difficile* diarrhea, leaving metronidazole as the mainstay of treatment in the United States where teicoplanin and fusidic acid are not marketed. It is likely that any new antimicrobial agent used to treat *C. difficile* will be similarly plagued by a high rate of recurrence, presumably incurred as a result of disruption of normal bowel flora. There is a need for improved treatment and prevention of this increasingly frequent and debilitating nosocomial infection. Treatments that utilize passive antibodies, immunization, nontoxigenic *C. difficile*, or other forms of biotherapy may hold the key to improved treatment and prevention of *C. difficile* disease in the future. In the meantime, it behooves all practitioners to use antimicrobials judiciously in order to prevent as many cases of *C. difficile* diarrhea as possible.

References

Altaie SS, Meyer P, Dryja D (1994) Comparison of two commercially available enzyme immunoassays for detection of *Clostridium difficile* in stool specimens. J Clin Microbiol 132:51–53

Arabi Y, Dimock F, Burdon DW, Alexander-Willliams J, Keighley MRB (1979) Influence of neomycin and metronidazole on colonic microflora of volunteers. J Antimicrob Chemother 5:531–537

Arrow SA, Croese L, Bowman RA, Riley TV (1994) Evaluation of three commercial enzyme immunoassay kits for detecting faecal *Clostridium difficile* toxins. J Clin Pathol 47:954–956

Bacon AE, McGrath S, Fekety R, Holloway WJ (1991) In vitro synergy studies with *Clostridium difficile*. Antimicrob Agents Chemother 35:582–583

Bannatyne R, Jackowski J (1987) Susceptibility of *Clostridium difficile* to metronidazole, its bioactive metabolites and tinidazole. Eur J Clin Microbiol 6:505

Bartlett JG (1992) The 10 most common questions about *Clostridium difficile*-associated diarrhea/colitis. Infect Dis Clin Pract 1:254–259

Bassetti S, Frei R, Zimmerli W (1998) Fungemia with *Saccharomyces cerevesiae* after treatment with *Saccharomyces boulardii*. Am J Med 105:71–72

Biavasco F, Manso E, Varaldo PE (1991) In vitro activities of ramoplanin and four glycopeptide antibiotics against clinical isolates of *Clostridium difficile*. Antimicrob Agents Chemother 35:195–197

Bolton RP, Culshaw MA (1986) Fecal metronidazole concentrations during oral and intravenous therapy for antibiotic-associated colitis due to *Clostridium difficile*. Gut 27:1169–1172

Bradbury AW, Barrett S (1997) Surgical aspects of *Clostridium difficile* colitis. Br J Surg 84:150–159

Buggy BP, Fekety R, Silva J Jr (1987) Therapy of relapsing *Clostridium difficile*-associated diarrhea and colitis with the combination of vancomycin and rifampin. J Clin Gastroenterol 9:155–159

Burdon DW, Brown DJ, Youngs DJ, Arabi Y, Shinagawa N, Alexander-Williams J, Keighley MRB (1979) Antibiotic susceptibility of *Clostridium difficile*. J Antimicrob Chemother 5:307–310

Chatila W, Manthous CA (1995) *Clostridium difficile* causing sepsis and an acute abdomen in critically ill patients. Crit Care Med 23:1146–1150

deLalla F, Nicolin R, Rinaldi E, Scarpellini P, Rigoli R, Manfrin V, Tramarin A (1992) Prospective study of oral teicoplanin versus oral vancomycin for therapy of pseudomembranous colitis and *Clostridium difficile*-associated diarrhea. Antimicrob Agents Chemother 36:2192–2196

Dudley MN, McLaughlin JC, Carrington G, Frick J, Nightingale CH, Quintiliani R (1986) Oral bacitracin versus vancomycin therapy for *Clostridium difficile*-induced diarrhea: A randomized, double-blind trial. Arch Intern Med 146:1101–1104

Elmer GW, Surawicz CM, McFarland LV (1996) Biotherapeutic Agents. J Amer Med Assoc 275:870–876

Fekety R (1997) Guidelines for the diagnosis and management of *Clostridium difficile*-associated diarrhea and colitis. Am J Gastroent 92:739–750

Fekety R, Shah AB (1993) Diagnosis and treatment of *Clostridium difficile* colitis. J Amer Med Assoc 269:71–75

Fekety R, Silva J, Kauffman C, Buggy B, Deery G (1989) Treatment of antibiotic-associated *Clostridium difficile* colitis with oral vancomycin: comparison of two dosage regimens. Am J Med 86:15–19

Finegold SM, George WL (1988) Therapy directed against *Clostridium difficile* and its toxins: complications of therapy. In: Rolfe RD, George WL (eds) *Clostridium difficile*: its role in intestinal disease. Academic Press, New York, pp 341–357

Gerding DN, Johnson S, Peterson LR, Mulligan ME, Silva J Jr (1995) *Clostridium difficile*-associated diarrhea and colitis. Infect Cont Hosp Epidemiol 16:459–477

Gerding DN, Olson MM, Peterson LR, Teasley DG, Gebhard RL, Schwartz ML, Lee JT Jr (1986) *Clostridium difficile*-associated diarrhea and colitis in adults: A prospective case-controlled epidemiologic study. Arch Intern Med 146:95–100

Gorbach SL (1996) *Clostridium difficile*-associated diarrhea. Infect Dis Clin Practice 5:84

Gorbach SL, Chang T-W, Goldin B (1987) Successful treatment of relapsing *Clostridium difficile* colitis with lactobacillus GG. Lancet ii:1519

Grundfest-Broniatowski S, Quader M, Alexander F, Walsh RM, Lavery I, Milsom J (1996) *Clostridium difficile* colitis in the critically ill. Dis Colon Rectum 39:619–623

Guzman R, Kirkpatrick J, Forward K, Lim F (1988) Failure of parenteral metronidazole in the treatment of pseudomembranous colitis. J Infect Dis 158:1146

Hoverstad T, Carlstedt-Duke B, Lingaas E, Miltvedt T, Norin KE, Saxerholt H, Steinbakk M (1986) Influence of ampicillin, clindamycin, and metronidazole on faecal excretion of short-chain fatty acids in healthy subjects. Scand J Gastronenterol 21:621–626

Hospital Infection Control Practices Advisory Committee (HICPAC) (1995) Recommendations for preventing the spread of vancomycin resistance. Infect Control Hospital Epidemiol 16:105–113

Ings RM, McFadzean JA, Ormerod WE (1975) The fate of metronidazole and its implications in chemotherapy. Xenobiotica 5:223–235

Jang SS, Hansen LM, Breher JE, Riley DA, Magdesian KG, Tang YJ, Silva J Jr (1997) Antimicrobial susceptibilities of equine isolates of *Clostridium difficile* and molecular characterization of metronidazole-resistant strains. Clin Infect Dis 25(Suppl 2):S266–277

Johnson S, Adelmann A, Clabots CR, Peterson LR, Gerding DN (1989) Recurrences of *Clostridium difficile* diarrhea not caused by the original infecting organism. J Infect Dis 159:340–343

Johnson S, Homann SR, Bettin KM, Quick JN, Clabots CR, Peterson LR, Gerding DN (1992) Treatment of asymptomatic *Clostridium difficile* carriers (fecal excretors) with vancomycin or metronidazole. Ann Intern Med 117:297–302

Keighley MR, Burdon DW, Arabi Y, Alexander-Williams J, Thompson H, Youngs D, Johnson M, Bentley S, George RH, Mogg GAG (1978) Randomized controlled trial of vancomycin for pseudomembranous colitis and postoperative diarrhoea. Br Med J 2:1667–1669

Kleinfeld DI, Sharpe RJ, Donta ST (1988) Parenteral therapy for antibiotic-associated pseudomembranous colitis. J Infect Dis 157:389

Leung DYM, Kelly CP, Boguniewicz M, Pothoulakis C, LaMont JT, Flores A (1991) Treatment with intravenously administered gamma globulin of chronic relapsing colitis induced by *Clostridium difficile* toxin. J Pediatr 118:633–637

Levett PN (1991) Time-dependent killing of *Clostridium difficile* by metronidazole and vancomycin. J Antimicrob Chemother 27:55–62

Liacouras CA, Piccoli DA (1996) Whole-bowel irrigation as an adjunct to the treatment of chronic, relapsing *Clostridium difficile* colitis. J Clin Gastroenterol 22:186–189

Lippsett PA, Samantaray DK, Tam ML, Bartlett JG, Lillemoe KD (1994) Pseudomembranous colitis: A surgical disease? Surgery 116:491–496

Lowenkron SE, Waxner J, Khullar P, Ilowite JS, Niederman MS, Fein AM (1996) *Clostridium difficile* infection as a cause of severe sepsis. Intensive Care Med 22:990

McFarland LV, Surawicz CM, Greenberg RN, Fekety R, Elmer G, Moyer KA, Melcher SA, Bowen KE, Cox JL, Noorani Z, Harrington G, Rubin M, Greenwald D (1994) Randomized placebo-controlled trial of *Saccharomyces boulardii* in combination with standard antibiotics for *Clostridium difficile* disease. J Amer Med Assoc 271:1913–1918

Mogg GAG, George RH, Youngs D, Johnson M, Thompson H, Burdon DW, Keighley MRB (1982) Randomized controlled trial of colestipol in antibiotic-associated colitis. Brit J Surg 69:137–139

Moncino MD, Falletta JM (1992) Multiple relapses of *Clostridium difficile*-associated diarrhea in a cancer patient: successful control with long-term cholestyramine therapy. Am J Pediatr Hematol Oncol 14:361–364

Morris JB, Zollinger RM, Stellato TA (1990) Role of surgery in antibiotic-induced pseudomembranous colitis. Am J Surg 160:535–539

Novak E, Lee JG, Seckman CE, Philips JP, DiSanto AR (1976) Unfavorable effect of atropine-diphenoxylate (Lomotil) therapy in lincomycin-caused diarrhea. J Amer Med Assoc 235:1451–1454

Olson MM, Shanholtzer MT, Lee JT Jr, Gerding DN (1994) Ten years of prospective *Clostridium difficile*-associated disease surveillance and treatment at the Minneapolis VA Medical Center, 1982–1991. Infect Cont Hosp Epidemiol 15:371–381

O'Neill GL, Beaman MH, Riley TV (1991) Relapse versus reinfection with *Clostridium difficile*. Epidemiol Infect 107:627–635

Panis Y, Hautefeuille P, Hecht Y, Le Houelleur J, Gompel H (1992) Surgical treatment of toxic megacolon complicating pseudomembranous colitis. Apropos of a case, review of the literature. Ann Chir 46:453–457

Peterson LR, Gerding DN (1990) Antimicrobial agents in *Clostridium difficile*-associated intestinal disease. In: Rambaud J-C, Ducluzeau R (eds) *Clostridium difficile*-associated intestinal diseases. Springer-Verlag, Paris, pp 115–127

Pruksananonda P, Powell KR (1989) Multiple relapses of *Clostridium difficile*-associated diarrhea responding to an extended course of cholestyramine. Pediatr Infect Dis J 8:175–178

Sanchez JL, Gerding DN, Olson MM, Johnson S (1999) Metronidazole susceptibility in *Clostridium difficile* isolates recovered from cases of *C. difficile*-associated disease treatment failures and successes. Anaerobe (in press)

Seal D, Borriello SP, Barclay F, Welch A, Piper M, Bonnycastle M (1987) Treatment of relapsing *Clostridium difficile* diarrhea by administration of a non-toxigenic strain. Eur J Clin Microbiol 6: 51–53

Silva J Jr (1989) Update on pseudomembranous colitis. West J Med 151:644–648

Silva J Jr, Batts DH, Fekety R, Plouffe JF, Rifkin GD, Baird I (1981) Treatment of *Clostridium difficile* colitis and diarrhea with vancomycin. Am J Med 71:815–822

Surawicz CM, McFarland LV, Elmer G, Chin J (1989) Treatment of recurrent *Clostridium difficile* colitis with vancomycin and *Saccharomyces boulardii*. Am J Gastroenterol 84:1285–1287

Swedish CDAD Study Group (1994) Treatment of *Clostridium difficile* associated diarrhea and colitis with an oral preparation of teicoplanin; a dose finding study. Scand J Infect Dis 26:309–316

Teasley DG, Gerding DN, Olson MM, Peterson LR, Gebhard RL, Schwartz ML, Lee ML Jr (1983) Prospective randomized trial of metronidazole versus vancomycin for *Clostridium difficile*-associated diarrhea and colitis. Lancet ii:1043–1046

Tedesco FJ, Gordon D, Fortson WC (1985) Approach to patients with multiple relapses of antibiotic-associated pseudomembranous colitis. Am J Gastroenterol 80:867–868

Triadafilopoulos G, Hallstone AE (1991) Acute abdomen as the first presentation of pseudomembranous colitis. Gastroenterology 101:685–691

Trudel JL, Deschenes M, Mayrand S, Barkun AN (1995) Toxic megacolon complicating pseudomembranous enterocolitis. Dis Colon Rectum 38:1033–1038

Tvede M, Rask-Madsen J (1989) Bacteriotherapy for chronic relapsing *Clostridium difficile* diarrhea in six patients. Lancet i:1156–1160

Warny M, Denie C, Delmee M, Lefebvre C (1995) Gamma globulin administration in relapsing *Clostridium difficile*-induced pseudomembranous colitis with a defective antibody response to toxin A. Acta Clin Belg 50:36–39

Wenisch C, Parschalk B, Hasenhundl M, Hirschl AM, Graninger W (1996) Comparison of vancomycin, teicoplanin, metronidazole, and fusidic acid for the treatment of *Clostridium difficile*-associated diarrhea. Clin Infect Dis 22:813–818

Wilcox MH (1998) Treatment of *Clostridium difficile* infection. J Antimicrob Chemother 41(Suppl C): 41–46

Wilcox MH, Fawley WN, Settle CD, Davidson A (1998) Recurrence of symptoms in *Clostridium difficile* infection–relapse or reinfection? J Hosp Infect 38:93–100

Wilcox MH, Howe R (1995) Diarrhoea caused by *Clostridium difficile*: response time for treatment with metronidazole and vancomycin. J Antimicrob Chemother 36:673–679

Young GP, Ward PB, Bayley N, Gordon D, Higgins G, Trapani JA, McDonald, MI, Labrooy J, Hecke R (1985) Antibiotic-associated colitis due to *Clostridium difficile*: Double-blind comparison of vancomycin with bacitracin. Gastroenterol 89:1038–1045

Zimmerman MJ, Bak A, Sutherland LR (1997) Review article: Treatment of *Clostridium difficile* infection. Aliment Pharmacol Ther 11:1003–1012

Zuckerman E, Kanel G, Ha C, Kahn J, Gottesman B-S, Korula J (1997) Low albumin gradient ascites complicating severe pseudomembranous colitis. Gastroenterology 112:991–994

Subject Index

Current Topics in Microbiology and Immunology

Volumes published since 1989 (and still available)

Vol. 209: **Miller, Virginia L. (Ed.):** Bacterial Invasiveness. 1996. 16 figs. IX, 115 pp. ISBN 3-540-60065-5

Vol. 210: **Potter, Michael; Rose, Noel R. (Eds.):** Immunology of Silicones. 1996. 136 figs. XX, 430 pp. ISBN 3-540-60272-0

Vol. 211: **Wolff, Linda; Perkins, Archibald S. (Eds.):** Molecular Aspects of Myeloid Stem Cell Development. 1996. 98 figs. XIV, 298 pp. ISBN 3-540-60414-6

Vol. 212: **Vainio, Olli; Imhof, Beat A. (Eds.):** Immunology and Developmental Biology of the Chicken. 1996. 43 figs. IX, 281 pp. ISBN 3-540-60585-1

Vol. 213/I: **Günthert, Ursula; Birchmeier, Walter (Eds.):** Attempts to Understand Metastasis Formation I. 1996. 35 figs. XV, 293 pp. ISBN 3-540-60680-7

Vol. 213/II: **Günthert, Ursula; Birchmeier, Walter (Eds.):** Attempts to Understand Metastasis Formation II. 1996. 33 figs. XV, 288 pp. ISBN 3-540-60681-5

Vol. 213/III: **Günthert, Ursula; Schlag, Peter M.; Birchmeier, Walter (Eds.):** Attempts to Understand Metastasis Formation III. 1996. 14 figs. XV, 262 pp. ISBN 3-540-60682-3

Vol. 214: **Kräusslich, Hans-Georg (Ed.):** Morphogenesis and Maturation of Retroviruses. 1996. 34 figs. XI, 344 pp. ISBN 3-540-60928-8

Vol. 215: **Shinnick, Thomas M. (Ed.):** Tuberculosis. 1996. 46 figs. XI, 307 pp. ISBN 3-540-60985-7

Vol. 216: **Rietschel, Ernst Th.; Wagner, Hermann (Eds.):** Pathology of Septic Shock. 1996. 34 figs. X, 321 pp. ISBN 3-540-61026-X

Vol. 217: **Jessberger, Rolf; Lieber, Michael R. (Eds.):** Molecular Analysis of DNA Rearrangements in the Immune System. 1996. 43 figs. IX, 224 pp. ISBN 3-540-61037-5

Vol. 218: **Berns, Kenneth I.; Giraud, Catherine (Eds.):** Adeno-Associated Virus (AAV) Vectors in Gene Therapy. 1996. 38 figs. IX,173 pp. ISBN 3-540-61076-6

Vol. 219: **Gross, Uwe (Ed.):** Toxoplasma gondii. 1996. 31 figs. XI, 274 pp. ISBN 3-540-61300-5

Vol. 220: **Rauscher, Frank J. III; Vogt, Peter K. (Eds.):** Chromosomal Translocations and Oncogenic Transcription Factors. 1997. 28 figs. XI, 166 pp. ISBN 3-540-61402-8

Vol. 221: **Kastan, Michael B. (Ed.):** Genetic Instability and Tumorigenesis. 1997. 12 figs.VII, 180 pp. ISBN 3-540-61518-0

Vol. 222: **Olding, Lars B. (Ed.):** Reproductive Immunology. 1997. 17 figs. XII, 219 pp. ISBN 3-540-61888-0

Vol. 223: **Tracy, S.; Chapman, N. M.; Mahy, B. W. J. (Eds.):** The Coxsackie B Viruses. 1997. 37 figs. VIII, 336 pp. ISBN 3-540-62390-6

Vol. 224: **Potter, Michael; Melchers, Fritz (Eds.):** C-Myc in B-Cell Neoplasia. 1997. 94 figs. XII, 291 pp. ISBN 3-540-62892-4

Vol. 225: **Vogt, Peter K.; Mahan, Michael J. (Eds.):** Bacterial Infection: Close Encounters at the Host Pathogen Interface. 1998. 15 figs. IX, 169 pp. ISBN 3-540-63260-3

Vol. 226: **Koprowski, Hilary; Weiner, David B. (Eds.):** DNA Vaccination/Genetic Vaccination. 1998. 31 figs. XVIII, 198 pp. ISBN 3-540-63392-8

Vol. 227: **Vogt, Peter K.; Reed, Steven I. (Eds.):** Cyclin Dependent Kinase (CDK) Inhibitors. 1998. 15 figs. XII, 169 pp. ISBN 3-540-63429-0

Vol. 228: **Pawson, Anthony I. (Ed.):** Protein Modules in Signal Transduction. 1998. 42 figs. IX, 368 pp. ISBN 3-540-63396-0

Vol. 229: **Kelsoe, Garnett; Flajnik, Martin (Eds.):** Somatic Diversification of Immune Responses. 1998. 38 figs. IX, 221 pp. ISBN 3-540-63608-0

Vol. 230: **Kärre, Klas; Colonna, Marco (Eds.):** Specificity, Function, and Development of NK Cells. 1998. 22 figs. IX, 248 pp. ISBN 3-540-63941-1

Vol. 231: **Holzmann, Bernhard; Wagner, Hermann (Eds.):** Leukocyte Integrins in the Immune System and Malignant Disease. 1998. 40 figs. XIII, 189 pp. ISBN 3-540-63609-9

Vol. 232: **Whitton, J. Lindsay (Ed.):** Antigen Presentation. 1998. 11 figs. IX, 244 pp. ISBN 3-540-63813-X

Vol. 233/I: **Tyler, Kenneth L.; Oldstone, Michael B. A. (Eds.):** Reoviruses I. 1998. 29 figs. XVIII, 223 pp. ISBN 3-540-63946-2

Vol. 233/II: **Tyler, Kenneth L.; Oldstone, Michael B. A. (Eds.):** Reoviruses II. 1998. 45 figs. XVI, 187 pp. ISBN 3-540-63947-0

Vol. 234: **Frankel, Arthur E. (Ed.):** Clinical Applications of Immunotoxins. 1999. 16 figs. IX, 122 pp. ISBN 3-540-64097-5

Vol. 235: **Klenk, Hans-Dieter (Ed.):** Marburg and Ebola Viruses. 1999. 34 figs. XI, 225 pp. ISBN 3-540-64729-5

Vol. 236: **Kraehenbuhl, Jean-Pierre; Neutra, Marian R. (Eds.):** Defense of Mucosal Surfaces: Pathogenesis, Immunity and Vaccines. 1999. 30 figs. IX, 296 pp. ISBN 3-540-64730-9

Vol. 237: **Claesson-Welsh, Lena (Ed.):** Vascular Growth Factors and Angiogenesis. 1999. 36 figs. X, 189 pp. ISBN 3-540-64731-7

Vol. 238: **Coffman, Robert L.; Romagnani, Sergio (Eds.):** Redirection of Th1 and Th2 Responses. 1999. 6 figs. IX, 148 pp. ISBN 3-540-65048-2

Vol. 239: **Vogt, Peter K.; Jackson, Andrew O. (Eds.):** Satellites and Defective Viral RNAs. 1999. 39 figs. XVI, 179 pp. ISBN 3-540-65049-0

Vol. 240: **Hammond, John; McGarvey, Peter; Yusibov, Vidadi (Eds.):** Plant Biotechnology. 1999. 12 figs. XII, 196 pp. ISBN 3-540-65104-7

Vol. 241: **Westblom, Tore U.; Czinn, Steven J.; Nedrud, John G. (Eds.):** Gastroduodenal Disease and Helicobacter pylori. 1999. 35 figs. XI, 313 pp. ISBN 3-540-65084-9

Vol. 242: **Hagedorn, Curt H.; Rice, Charles M. (Eds.):** The Hepatitis C Viruses. 2000. 47 figs. IX, 379 pp. ISBN 3-540-65358-9

Vol. 243: **Famulok, Michael; Winnacker, Ernst-L.; Wong, Chi-Huey (Eds.):** Combinatorial Chemistry in Biology. 1999. 48 figs. IX, 189 pp. ISBN 3-540-65704-5

Vol. 244: **Daëron, Marc; Vivier, Eric (Eds.):** Immunoreceptor Tyrosine-Based Inhibition Motifs. 1999. 20 figs. VIII, 179 pp. ISBN 3-540-65789-4

Vol. 245/I: **Justement, Louis B.; Siminovitch, Katherine A. (Eds.):** Signal Transduction and the Coordination of B Lymphocyte Development and Function I. 2000. 22 figs. XVI, 274 pp. ISBN 3-540-66002-X

Vol. 245/II: **Justement, Louis B.; Siminovitch, Katherine A. (Eds.):** Signal Transduction on the Coordination of B Lymphocyte Development and Function II. 2000. 13 figs. XV, 172 pp. ISBN 3-540-66003-8

Vol. 246: **Melchers, Fritz; Potter, Michael (Eds.):** Mechanisms of B Cell Neoplasia 1998. 1999. 111 figs. XXIX, 415 pp. ISBN 3-540-65759-2

Vol. 247: **Wagner, Hermann (Ed.):** Immunobiology of Bacterial CpG-DNA. 2000. 34 figs. IX, 246 pp. ISBN 3-540-66400-9

Vol. 248: **du Pasquier, Louis; Litman, Gary W. (Eds.):** Origin and Evolution of the Vertebrate Immune System. 2000. 81 figs. IX, 324 pp. ISBN 3-540-66414-9

Vol. 249: **Jones, Peter A.; Vogt, Peter K. (Eds.):** DNA Methylation and Cancer. 2000. 16 figs. IX, 169 pp. ISBN 3-540-66608-7

Printing: Saladruck, Berlin
Binding: H. Stürtz AG, Würzburg